N. Smith R.
Siskiyou Mts.
Klamath R.
NORTH COAST RANGES
KLAMATH RANGES
Mt. Eddy
Salmon Mts.
Mt. Shasta
CASCADE RANGE
Lake Shasta
Pit R.
Modoc Plateau
Warner Mts.

Eureka
Humbolt Bay
Cape Mendocino

Eagle Lake

Mt. Lassen
Susanville

Mad R.
Trinity R.
Eel River
Russian R.
Noyo R.
Mendocino

Honey Lake

Quincy

Red Mt.
Clear Lake
Yollo Bolly Mts.
Sacramento R.

Chico
Lake Oroville
Oroville
Sutter Buttes
Truckee

Lake Tahoe

Point Reyes
Mt. Tamalpais

Santa Rosa
Lake Berryessa
GREAT VALLEY
Yuba R.
Folsom Lake
Sacramento
SIERRA
Carson Pass
Sonora Pass
Yosemite Valley

Mono Lake

San Francisco
San Mateo
Antioch
Mt. Diablo
Ione
Consumnes R.
Stanislaus R.
Jamestown
Tuolumne R.
NEVADA RANGE
Owens R.
WHITE INYO RANGE
Owens River

Bishop

Santa Clara
Santa Cruz
Monterey Bay
Monterey

SOUTH COAST RANGES
San Joaquin R.
Merced R.
San Jouquin R.
Fresno
Fresno Slough
Kings R.
Olancha Peak
Owens Lake

Death Valley

Big Sur
Salinas River
Santa Lucia Mts.
San Benito Mt.

Isabella Lake
Kern R.
Owens Peak

Mojave

New York Mts.
PIUTE RANGE
Colorado River

Morro Bay
San Luis Obisbo
Carrizo Plain
Mt. Pinos

Bakersfield
Red Rock Canyon

Mojave River
Desert

Santa Maria R.
Figueroa Mt.

San Gabriel Mts.
Big Bear Lake
San Bernadino Mts.

Santa Inez Mts.
TRANSVERSE RANGE

Santa Barbara
Santa Monica Mts.
Los Angeles
Palos Verdes
Upper Newport Bay
Santa Ana R.
Santa Ana Mts.
Santa Rosa Plateau
Vail Lake
Mt. Palomar
PENINSULAR RANGES
Laguna Mts.
San Jacinto Mt.
Anza Borrego
Salton Sea
Algodones Dunes

CHANNEL ISLANDS

Cuyamaca Peak
McGinty Mt.
Colorado Desert

San Diego
Otay Mt.

CALIFORNIA'S
WILD
GARDENS

A Guide to Favorite Botanical Sites

Edited by

PHYLLIS M. FABER

Published under the auspices of

THE CALIFORNIA NATIVE PLANT SOCIETY

and THE CALIFORNIA DEPARTMENT OF FISH AND GAME

UNIVERSITY OF CALIFORNIA PRESS

BERKELEY LOS ANGELES LONDON

CALIFORNIA'S
WILD
GARDENS

A Guide to Favorite Botanical Sites

Edited by

PHYLLIS M. FABER

Published under the auspices of

THE CALIFORNIA NATIVE PLANT SOCIETY
and THE CALIFORNIA DEPARTMENT OF FISH AND GAME

UNIVERSITY OF CALIFORNIA PRESS

BERKELEY LOS ANGELES LONDON

DEDICATION

This book is dedicated to those who work for the protection and preservation of California's native plants for this and future generations.

University of California Press
Berkeley and Los Angeles, California

University of California Press, Ltd.
London, England

Published under the auspices of
The California Department of Fish and Game and
The California Native Plant Society

© 1997 California Department of Fish and Game
2nd Printing 2005
1st printing 1997, published as:
   California's Wild Gardens: A Living Legacy

Printed in Hong Kong through Global Interprint, Petaluma, California.
This book was set in Bembo, Trajan, Castellar, and Geometrica Titling.

13 12 11 10 09 08 07 06 05 04
10 9 8 7 6 5 4 3 2 1

Book and cover design by Beth Hansen-Winter
Front cover photograph: Lupines and purple owl's-clover (*Lupinus* ssp. and *Castilleja exserta* ssp. *exserta*) by Craig Lippus; Back cover photographs (clockwise from top right): Bristlecone pine (*Pinus longaeva*) by B. and G. Corsi, giant sequoias (*Sequoiadendron giganteum*) by B. Evarts, flannelbush (*Fremontodendron californicum*) by W. and W. Follette, and Wallace's nightshade (*Solanum wallacei*) by L. Ulrich

Library of Congress Cataloging-in-Publication Data

California's wild gardens : a guide to favorite botanical sites / Phyllis M. Faber.
   p. cm.
   Originally published: California's wild gardens : a living legacy. Sacramento, CA : California Native Plant Society, 1997.
   Includes bibliographical references and index.
   ISBN 0-520-24031-6 (pbk. : alk. paper)
   1. Botany—California. 2. Phytogeography—California. 3. Plant ecophysiology—California. I. Faber, Phyllis M.
QK149.C3217 2005
581.9794—dc22

2004062037

PHOTO CREDITS

Legend: T: TOP; B: BOTTOM; L: LEFT; M: MIDDLE; R: RIGHT; U: UPPER; LW: LOWER

F. Almeda: 214B; **W. Anderson:** 2B, 18, 34, 42B, 45T, 108, 110, 111TR, 114TR, 114ML, 114B, 115B, 118TR, 118MR, 186BL; **W. Armstrong:** 216T; **M. Austin-McDermon:** 5T, 43TR, 51, 54T, 55T, 58TR, 58ML, 58BL, 76TR, 79BR, 83B, 84L, 85TR, 85MR, 91BL, 94ML, 102BL, 107TR, 113BL, 124BL, 128L, 128BM, 130BL, 137BL, 166TR, 172R, 178UL, 189UR, 193BR, 195TL, 205BL, 209MR; **B.C. Baldwin:** 98BR, 99B; **G. Ballmer:** 31M; **F. Balthis:** 14, 40, 86TR, 90TR, 90ML, 90B, 93T, 93T, 141B, 142BR, 155M, 169B, 162TR, 163; **M. Barbour:** 27B, 29T; **K. Berg:** 35; **R. Bittman:** 131TR; **T. Ansel Blake:** 59, 107B, 116; **D. Briggs:** 113UR; **J. Briggs:** 97TR; **A. Brinkmann-Busi:** 165BL; **D. Cavagnaro:** 79TR, 103BR, 119BR; **D. Cheatham:** 102T, 105BR, 130TL, 136, 149TL, 157TR, 159T, 168BL; **G. Clark:** 128BR, 130M, 131B; **S. Cochrane:** 5M, 8, 9T, 9BL, 10M, 11TR, 12, 31B, 50B, 64, 65B, 101LR, 114T, 126TL, 127BR, 128TR, 133R, 138UL, 142T, 143T, 143MR, 146ML, 156BL, 159B, 160TL, 160BL, 160BM, 161TL, 161B, 162UL, 162BR, 163, 168MR, 171T, 173T, 174L, 194BR, 207L, 207BR, 212ML; **E. Cooper:** 61; **J. Cornett:** 217TR; **B. and G. Corsi:** 2T, 6, 30T, 32, 65TR, 65MR, 97B, 105TR, 105UMR, 109TL, 111MT, 111M, 111MR, 123UMR, 126ML, 152TR, 198TR, 198MR, 199TR, 199BM, 200TR, 203MR, 208TM, 208BL, 209M, 209BR, 211ML, 211BL, 211MB, 211TR, 211LwMR, 212TR; **K. Davis:** 112L; **T. Dodson:** 21, 29B, 158BL, 167TM, 195BR, 207ML; **D. Eastman:** 41T, 41B; **S. Edwards:** 53; **E. Ely:** 81BR; **B. Evarts:** 1, 17, 134TR, 188T, 188B, 202, 203B; **P. Faber:** 24B, 45B, 178BR; **G. Fellers:** 74BL; **W. and W. Follette:** 4T, 4B, 19, 25, 26, 28, 30B, In, In, 62L, 72T, 73BM, 92AL, 92BM, 92BR, 93BR, 99MR, 113LR, 117TR, 120BL, 123BL, 207TR, 213; **E. Frank:** 165BL; **J. Game:** 52L, 98T, 99TM, 99TR, 101TR; **D. Gilman:** 163, 196TR; **M. Graf:** 137BM, 139TR, 139B, 140BM, 141TM, 141TR, 143BR, 146BL, 146BR, 147ML, 147BL; **M.A. Griggs:** 118BL; **R. Gustafson:** 165TR; **J. Haas:** 144TR, 144BR; **J.R. Haller:** 67B, 94BL, 169TR; **J. Harris:** 52TR, 141TL, 147 R; **A. Hayler:** 88BL; **L. Heckard:** 55BL; **R. Herrman:** 153, 181TL; **D. Hillyard:** 84 R, 91TR, 92BL; **S. Holt:** iv-v, 55BL, 55BR, 70B, 82TR, 170BL, 193LwMR; **D. Imper:** 204T; **P. Johnson:** 214TL, 215BM; **T. Johnson:** 89T, 107TM; **J. Jokerst:** 66BL; **S. Junak:** 105BL, 152BL, 155TL, 155TR, 156TR, 156MR, 157TL, 164BL, 164BR, 165MR, 176BM; **T. Keeler-Wolf:** 48TR; **D. Keil:** 103TR; **T. Krantz:** 170TM, 173B; **C. Lippus:** cover, 3, 9BR, 22, 27TR, 31, 71TR, 80R, 95, 109BR, 121T, 121BM, 122, 134BL, 135BL, 135BM, 135BR, 201, 210; **L. Lozier:** 185MR; **H. Malde:** 187; **S. McKelvey:** 38LwL, 73TR, 73UMR, 73BR, 86L; **M. McLeod:** 106; **J. Medeiros:** 15, 117MR, 133TL; **C. Millar:** 56, 57R; **S. Montgomery:** 216BL, 217BR; **N. Nedeff:** 91MR, 94TR; **L. Norris:** 148BM, 151TR; **T. Oberbauer:** 157BL, 177T, 177B, 180BR, 181BR, 183BR, 189TL, 189B, 190T, 190BL, 191BL, 191TR, 191MR, 192BL, 192BR, 193TR, 193UMR, 194TR; **B. O'Brien:** 100B; **J. Ordano:** 16, 36, 44BL, 68, 69TR, 70T, 71BL, 76BR, 78B, 80L, 100TR, 109TR, 112BL, 117B, 134TL, 137TR, 200B; **R. Parsons:** 62TR, 71TL, 198BL; **B. Pavlik:** 82BR; **B. "M." Peterson:** 180TR; **A. Pickart:** 43BR; **R. Pickup:** 54B, 132B; **R. Raiche:** 49T, 49B; **E. Ross:** 43TL, 73LMR, 77TR, 77BL; **G. Rowell:** 72B; **D. Sanger:** 81T; **R. Schlising:** 111BL; **L. Serpa:** 37; **J. Shevock:** 144ML, 145TR, 145B, 148L, 149TR, 149B, 150, 151TM, 151MR, 151BM, 151BR; **M. Skinner:** 5B, 7, 10B, 11LwL, 13, 23T, 23M, 23B, 41M, 44TL, 50T, 57BL, 63BR, 69MR, 78T, 79MR, 83T, 85TL, 85BR, 86BR, 87, 102ML, 104TL, 104TR, 104B, 107MR, 123TR, 123BR, 125TL, 125BR, 129BL, 129BR, 132T, 133UML, 135TR, 138LwL, 139BR, 164TR, 168TR, 171BL, 171BR, 172L, 180BL, 184TR, 184ML, 184BL, 185BR, 194BL, 195ML, 195TR, 203TR, 204BL, 205TR, 211UMR, 212MR, 217TL; **L. Smith:** 24T; **J. Smith, Jr.:** 43ML; **J.M. Stewart:** 186UL; **K. Stockwell:** 175T, 175B, 182TR, 182B, 183TL, 184BR; **D. Suzio:** 76BL, 212B; **D. Thomas:** 166B, 167MR; **T. Thomas:** 206; **L. Ulrich:** iii, 10T, 11MR, 11LwR, 38T, 39, 42T, 46T, 46M, 46BL, 47T, 47B, 60L, 60BM, 62BR, 63T, 66TR, 66MR, 67T, 74T, 74M, 75, 89B, 96, 97ML, 127MR, 140BL, 154, 155TR, 162LwL, 176BL, 179T, 179BL, 179BR, 196TR, 196BL, 197, 209T, 215TL, 215TR, 215BR; **R. York:** 48TL, 152TL; **G. Zahm:** 112BR, 121BL, 121BR, 124BR

Acknowledgments

When this project was first discussed in 1992 with Ken Berg, then Endangered Plant Program Coordinator for the Department of Fish and Game, I thought it might take a year or so to complete. Because of my years as editor of the California Native Plant Society's journal, *Fremontia*, I knew who and where various people were who were familiar with a list of special sites around California. Each would be asked to contribute short pieces about their favorite places and a book would thus be born. The authors did indeed generously donate their time and knowledge to the project; however, it is now more than five years later. The process was more complex and took far longer than I ever dreamed. Along the way, however, a number of people helped in important ways and I thank each of them.

Susan Cochrane, Chief of the Natural Heritage Division of the Department of Fish and Game, and her staff, Sandra Morey and Diane Ikeda of the Plant Conservation Program, were supportive and helpful in all phases of the project. Roxanne Bittman of the Division's Natural Diversity Data Base was invaluable with her vast knowledge of species distribution and her review of the slides.

Several people helped with the editorial process during different phases of the project. Early on, Crawford Cooley, Barbara Leitner, Nora Harlow, Robert Ornduff reviewed and made suggestions on initial submissions. Keith Howell, Gordie Slack, and Blake Edgar from the publications department of the California Academy of Sciences and Jon Stewart from the San Francisco Chronicle made sure the writing was understandable for a lay reader, and Nora Harlow proofread the manuscript on several occasions. Plant ecologists in the Department of Fish and Game, Gene Cooley, James Dice, Deborah Hillyard, Julie Horenstein, Ann Howald, Todd Keeler-Wolf, Richard Lis, Mary Meyer, Craig Martz, James Nelson, and Julie Vanderweir reviewed the manuscript, and Claudia Buck and Jeanne Clark provided editorial assistance. Mark Skinner, botanist with CNPS, reviewed the manuscript, checked photo plant identifications and wrote some of the captions. Regretfully, a small number of pieces were dropped because of a lack of photographic material. My apologies to these authors.

Frank Almeda and Tom Daniel from the Botany Department of the California Academy of Sciences reviewed both slides and the manuscript. Tom White, now at the University of California Press, provided ongoing encouragement as well as a preliminary marketing strategy for the book. Paula Nelson prepared the maps and Trisha Lamb Feuerstein the index.

The photographers deserve a special word of gratitude. Because the book project shifted during its course and the course was so much longer than anyone anticipated, I held some of their submissions for over four years. This was exceedingly frustrating, particularly for the professional photographers. That we have so many of their photographs in the book speaks to their generosity and goodwill.

In addition to her devoted attention to the design of this book, Beth Hansen, our designer extraordinaire, spent literally hundreds of hours pouring over slide submissions to select the best for each site described. This was uniquely difficult since botanists tend to take photographs when they are visiting a site regardless of weather or time of day (these slides are usually well identified but are often out of focus or gloomy), whereas professionals submit gorgeous images that are often unidentified as to species or site taken. Christy Sloan maintained computer records for all the hundreds of slides submitted.

Funding for this project was received from the California Department of Fish and Game, the U.S. Department of Defense, the U.S. Army Corps of Engineers (Sacramento and Los Angeles Districts), the U.S. Forest Service, and the U.S. Fish and Wildlife Service (Sacramento Office).

This book is the product of many contributors who gave freely of their time and talents towards a goal of preserving the unique and rich flora of California.

Phyllis M. Faber, Editor

Contents

N

PREFACE

California is blessed with a wealth of plant life that reflects the diversity of its natural landscape. Showcasing some of the state's rare and unique plants and their distinctive habitats along with some of the more common treasures, this book attempts to provide an introduction to California's priceless botanical heritage.

California is viewed as a series of ecological regions, each harboring a distinctive assemblage of plants specifically adapted to the prevailing environmental conditions. Within these regions are smaller areas, equivalent to a birder's "hot-spots" or the rockhound's gem areas, where local environmental factors have favored a special ensemble of rare or endemic plants. These botanical "hot-spots" are the focus of this book—in essence, a compendium of some of the floristically most important sites in our state.

Although these local spots of botanical interest are of immeasurable natural value, they often coincide with areas of great or potential commercial value. Over the past century California's natural landscapes have undergone dramatic changes as agriculture has replaced valley lands, forests have been logged, and millions of people have moved their homes and livelihoods here. As a result, many of the plant species unique to this state are now threatened with extinction. In recent decades, Californians have set a national example in protecting the state's resources, botanical and otherwise. However we continue to face many difficult decisions in our efforts to protect the state's diverse natural resources. To this end, this book provides examples of some of the botanical resources that we have, threats those resources face, and efforts underway to protect both plants and their habitats.

This book is written like the best travel guides, enticing the reader to visit and observe plants and their habitats. In its authorship, the book demonstrates the diversity of plant enthusiasts in California. More than 100 botanists from many different professional arenas have contributed articles. Their commitment to this project underscores the cooperative spirit that we feel is necessary for the long term conservation of botanical diversity in California.

It is with great pride that we have come together to produce this beautiful and educational resource. Each of the organizations we represent contributes uniquely to plant conservation. The California Department of Fish and Game is the trustee agency for the state's natural resources with responsibility to manage the state's diverse fish, wildlife, and plant resources, and the habitats upon which they depend for their ecological values and for their use and enjoyment by the public. The California Native Plant Society has a commitment to preserve the state's native flora through a grass-roots nonprofit organization that maintains a data base for California's rare plants and educates the public through field trips and publications. The California Academy of Sciences is a private research and educational institution with a focus on the systematic documentation and public awareness of our state's biological diversity.

By highlighting some of California's botanical bounty, our hope is that this book will inspire further exploration and discovery and that it will instill an appreciation for the necessity of preserving our natural heritage for generations to come.

Jacqueline E. Schafer, Director
California Department of Fish and Game

Lori Hubbart, President
California Native Plant Society

Dr. Evelyn Handler, Director
California Academy of Sciences

WHY IS CALIFORNIA'S FLORA SO RICH?

More than 6,000 species, subspecies, and varieties of native flowering plants, conifers, and ferns grow in the gentle oak woodlands, lofty mountains, spacious deserts, and along the magnificent coast of California. This is nearly one-fourth of all the plant types found in North America north of the Mexican border and more than are found in any other state. In an area comparable in size, all of New England has fewer than 2,000 plant species. Of the other states, only the huge expanse of Texas has more than 5,000 native plants.

But these impressive numbers alone do not tell the full story of California's rich flora. Our coast redwood (*Sequoia sempervirens*) is the tallest tree in existence with some individual trees growing to nearly 370 feet. Visitors from around the world marvel at the solemn splendor of the coast redwood, and at the immensity of the Sierra Nevada's giant sequoia (*Sequoiadendron giganteum*), the coast redwood's first cousin and the largest living being to grace the earth. And on high slopes of the arid White Mountains, the western bristlecone pine (*Pinus longaeva*) includes some of the oldest individuals of any species alive today, a few nearly 5,000 years of age. In a land of great contrasts, California is also home to the smallest flowering plant in existence, the pond-dwelling water-meal (*Wolffia globosa*). The entire plant is only a fraction of a tenth of an inch.

Why is our flora so rich, and why are so many plants found here and no place else on earth? The answer lies in part in California's complex geologic history and in part in its diverse topography, soils, and climate.

Fall brings color to the understory of giant sequoias (Sequoiadendron giganteum) *in Nelder Grove in the Sierra National Forest.*

GEOLOGIC HISTORY

When dinosaurs roamed the earth over sixty million years ago, California's central valley was the western edge of the North American continent. Major activity along three tectonic plates floating on the earth's molten core changed all that. The oceanic Farallon plate moved north, dipping under the lighter North American plate, and was simultaneously crushed and partially obliterated by the enormous Pacific plate to the west. This geologic upheaval lifted the Sierra Nevada, spawned repeated episodes of volcanism in the Cascade Mountains, and made Mount Lassen and Mount Shasta the massive volcanoes they are today. Unrelated rock masses, each with unique geological origins and composed of a host of materials with different chemistries, collided and joined along California's coast. Upwelling molten rock cooled to form bands of serpentine, gabbro, and other rocks, which eroded into soils that are toxic to many plants. Thus was born a highly complex geologic mosaic of rock types that break down into a rich and diverse assortment of soil types. These, in turn, shape the distribution of California's flora.

Massive outcrops of limestone in the White Mountains, BELOW, *decay into suitable soil and habitat for the long-lived bristlecone pine (Pinus longaeva). • A twenty-three million year old lava flow,* BOTTOM, *now exposed in these fragmenting vertical cliffs at Table Mountain in Butte County, is part of an early Oligocene basaltic formation.*

DIVERSE SOILS

California has all eleven of the world's major soil groups and ten percent of all named soils in the United States—approximately 1,200 different soil types. Some of these soils are as much as 500,000 years old. Some are high in toxic elements, low in essential nutrients, or extremely acidic. Some of these, such as serpentine and gabbro, are toxic to many plant species. Some of California's rarest plants are edaphic endemics, meaning they grow only on particular soil types. These include the picturesque twisted Sierra juniper (*Juniperus occidentalis* var. *australis*) that grows on solid granite outcrops; massive valley oak (*Quercus lobata*) in the Great Valley that grow on deep, well drained, structureless sediment within thirty feet of groundwater; and certain stands of ghost-like gray pine (*Pinus sabiniana*) on island outcrops of chromium- and nickel-laden serpentine soil amidst a sea of foothill brush. Isolated patches of serpentine and other unusual soils continue to provide California botanists with exciting discoveries.

SOIL TYPE PREFERENCES OF RARE SPECIES

The total number of rare plant species in California is 1,742.

SOIL TYPE	NUMBER OF RARE SPECIES
Serpentine	285
Granite	109
Clay	94
Carbonate	90
Volcanic	88
Alkaline	62
Gabbro	20
Sandstone	16
Shale	10
Gypsum	1

Data are updated from the CNPS *Inventory of Rare and Endangered Vascular Plants of California* (Skinner and Pavlik, 1994).

SERPENTINE AND ITS PLANT LIFE IN CALIFORNIA

California has the largest exposure of serpentine in North America. This unusual family of rocks, technically called ultramafic or serpentinite and commonly called serpentine, is derived from the earth's mantle and the surface where oceanic and continental plates collide. Serpentine rocks, ranging in color from dark red to shiny gray, green, or black, have in common iron magnesium silicate and impurities of chromium, nickel, and other toxic metallic elements.

As these rocks weather, soils develop that are high in magnesium, low in calcium, and toxic to most plants. Dramatic contrasts in vegetation occur; for example, chaparral or grassland communities may develop on serpentine soils, while either forest or shrubland may flourish on other soils nearby.

Major serpentine rock outcrops occur in the Klamath Mountains and in the inner and outer Coast Ranges, particularly in the North Bay counties of Napa, Sonoma, Mendocino, and Lake, the San Francisco Bay Area on Mount Tamalpais and the Tiburon Peninsula, the San Francisco Peninsula south to San Jose and Morgan Hill, and in San Benito County. The western Sierra Nevada also has major outcrops in narrow parallel bands running in a roughly north to south direction, extending from Tulare County north to Plumas County and including the Red Hills area of Tuolumne County near Chinese Camp. Its red soil and distinctive chaparral woodland vegetation, including buck brush (*Ceanothus cuneatus*) and gray pine (*Pinus sabiniana*), are its trademarks.

Serpentine rock is often shiny gray-green, BELOW, *and toxic to many plants. • A serpentine outcrop occurs on Carson Ridge in Marin County,* BOTTOM. *California has the largest exposure of serpentine rock in North America.*

Plants have adapted to serpentine soils everywhere that serpentine rocks reach the surface of the planet. Such plants range from strict serpentine endemics, or those that grow only on serpentine, such as leather oak (*Quercus durata*), milkwort jewelflower (*Streptanthus polygaloides*), and rock phacelia (*Phacelia egena*), to species such as Jeffrey pine (*Pinus jeffreyi*) and incense cedar (*Calocedrus decurrens*) that are found on granite in the Sierra but grow largely on serpentine in the North Coast Ranges. Still other native plants may grow either on and off serpentine, but often take the form of genetically fixed serpentine-tolerant races when they encounter the demanding serpentine soil. Some serpentine endemics such as the milkwort jewel-flower not only tolerate the toxic nickel, but accumulate it.

Plants restricted to serpentine contribute impressively to the list of California endemics. Over 200 species, subspecies, and varieties are found only, or mostly, on serpentine soils. Two species of cypress, Sargent cypress (*Cupressus sargentii*) and Mcnab cypress (*C. macnabiana*), and a few species of onion (*Allium*), mariposa lily (*Calochortus*), and mission bells (*Fritillaria*), are serpentine endemics. Several plant families are adapted to growing on serpentine soils and show high degrees of endemism: mustard (especially the genera *Streptanthus* and *Arabis*); buckwheat (*Eriogonum*); parsley (mostly *Lomatium* and *Perideridia*); flax (nearly all species of *Hespero-linon*); figwort (*Mimulus, Collinsia, Castilleja*); waterleaf (primarily *Phacelia*); and several genera of the sunflower family.

For the most part, these serpentine endemics have close relatives that are not endemic to serpentine, suggesting that evolution of a serpentine endemic is traceable to species of the nearby flora. Individuals of a non-serpentine species may become genetically adapted to serpentine (tolerating low calcium, high magnesium, and overall low soil fertility) and then multiply to form edaphic races. Subsequent divergence in floral and vegetative features, along with reproductive isolation, may produce species endemic to serpentine. This scenario is illustrated in the genus *Streptanthus*, where the serpentine races of *S. glandulosus* may have served as ancestors to serpentine endemics such as San Benito jewelflower (*S. insignis*) or Tiburon jewelflower (*S. niger*).

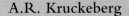

A.R. Kruckeberg

Three rare endemic plants occur only on serpentine soils: fountain thistle (Cirsium fontinale var. fontinale), TOP, known from four occurrences near Crystal Springs Reservoir in San Mateo County; Mount Cobb lupine (Lupinus sericatus), MIDDLE, in Lake County, threatened by geothermal development; and Tiburon jewelflower (Streptanthus niger), BOTTOM, occurring in only three populations in Marin County.

MANY CLIMATES AND MICROCLIMATES

Following the demise of the dinosaurs around sixty million years ago, California's climate cooled and changed from warm tropical to temperate. Forty million years ago, forests covered vast areas of the northern part of California, composed largely of temperate climate trees, both conifers and broadleaves, similar to those found in mild northern and eastern parts of North America today. This Arcto-Tertiary geoflora is one of three geofloras, or prehistorical floras based on geography and climate, that contribute to our current landscape. Pine (*Pinus*), spruce (*Picea*), alder (*Alnus*), and willow (*Salix*) were abundant. Herbaceous plants such as larkspur (*Delphinium*), buttercup (*Ranunculus*), sedge (especially *Carex*), onion (*Allium*), and iris (*Iris*) also occupied California as representatives of the Arcto-Tertiary geoflora, and continue to do so today.

Its odd thick trunk gives the elephant tree (Bursera microphylla) *its common name.*

Four million years ago (mid-Miocene epoch), California's landscape and climate were beginning to take shape. The San Andreas fault system marked the boundary of the northward moving Pacific plate and the stabilized North American Plate. Continued activity along this boundary built the Coast Ranges to the north and the Transverse and Peninsular ranges to the south. Cool ocean currents brought moisture to the California coastline in the form of fog, but elsewhere the climate was drying out. The seasonality of our Mediterranean climate developed, as the mountains created rainshadows and gradually dried the inland seas occupying the Sacramento and San Joaquin valleys.

As the climate warmed and dried, many elements of the Arcto-Tertiary geoflora retreated to higher elevations, to the north and to the coast, and were replaced by an invasion of more drought-tolerant, often hard-leaved (or sclerophyllous) plants from the south and east. This Madro-Tertiary geoflora came to occupy the drier lowlands in California, and mixed with the Arcto-Tertiary geoflora where the climate was cooler and more moist. Some of California's most familiar plants found west of the Sierra Nevada are part of the Madro-Tertiary flora, including live oak (*Quercus agrifolia*), which dominates coastal woodlands, California bay laurel (*Umbellularia californica*), madrone (*Arbutus menziesii*), and woody shrubs such as manzanita (*Arctostaphylos*), California lilac (*Ceanothus*), sumacs (*Rhus* spp. and *Malosma laurina*), and mountain-mahogany (*Cercocarpus*).

Between three and fifteen million years ago, during the Miocene and Pliocene epochs, lands that today are deserts were warmer and drier but not so arid as they are today. Madro-Tertiary sclerophylls, well suited to drought, shared space with remnant subtropicals in exotic families such as sapodilla (Sapotaceae), fig (Moraceae), and litchi (Sapindaceae), now long gone from the California flora.

As drying intensified over the past three million years, deserts expanded, bringing with them warm desert and temperate elements of the third major floristic immigration to California: the true xerophytes, or "dry plants," of the Colorado Desert. These are the prominent forms of the desert: cacti of all shapes and sizes, bizarre ocotillo (*Fouquieria splendens*), ferocious agaves (*Agave*) and yuccas (*Yucca*), thick-trunked elephant tree (*Bursera microphylla*), smoke trees (*Psorothamnus*), and graceful ironwood (*Olneya tesota*).

The red flowers of ocotillo (Fouquieria splendens) *are a frequent sight in the Colorado Desert in March.*

CALIFORNIA'S MEDITERRANEAN CLIMATE

Most of California has a Mediterranean climate, similar to that of lands bordering the Mediterranean Sea. Winters are cool and wet; summers are hot and dry. Hard frost and snow in winter are rare; summer temperatures may approach 100° F day after day. Most of the ten to thirty inches of annual precipitation falls as rain in the winter months of October through March. The growing season is short, beginning with the onset of winter rains and ending with summer drought. California's low-elevation landscapes typically are green in winter, brown in summer.

This type of climate is found in only three other parts of the world besides California and the Mediterranean rim: central Chile, the Cape region of South Africa, and southern and western parts of Australia. Certain types of vegetation—chaparral, foothill woodland, coastal scrub, montane evergreen forest—occur in every Mediterranean climate. A Californian visiting Spain, South Africa, Chile, or South Australia will see several familiar landscapes.

Along the coast there is little temperature change from day to night or from season to season. A cold, offshore current cools summer air, forming fog banks that shield the land and its vegetation from direct sunlight. Coastal climate is a maritime version of the Mediterranean climate. If one moves inland less than 100 miles, the moderating effect of the ocean is lost. Daily and seasonal oscillations of temperature become magnified.

Rainfall declines from north to south. Crescent City, in the northwest, enjoys six feet of rain a year, while San Francisco receives only twenty-five inches and San Diego, in the southwest, only ten. The length and intensity of the summer-dry period increases southward.

Both temperature and rainfall are affected by elevation. Temperatures fall about 3° F for every 1,000 foot rise in elevation. Annual precipitation (rain and snow) increases about seven inches for every 1,000 feet of rise up to 8,000 feet elevation. Above this, most of the moisture has been wrung out of the air and precipitation lessens.

Michael Barbour and Valerie Whitworth

FIRE AND ITS ROLE IN REJUVENATING PLANT COMMUNITIES

Wildfire can be a natural event in Mediterranean climates where lightning strikes are not always followed by rainfall. Until Californians adopted a fire-suppression policy eighty years ago, fire was a regularly expected natural event in many California vegetation types below 6,000 feet elevation, and the same acre of ground could be expected to burn every ten to fifty years. Fire was uncommon only in wetlands, deserts, and at high elevations.

In California, plants evolved over millions of years with fire as a natural environmental factor. As a result, many Californian species not only survive fire, but some require fire in order to complete their life cycle or to remain vigorous. Closed-cone conifer stands of pygmy cypress, Monterey cypress, and knobcone pine release few seeds from their tightly closed cones until a fire rages through, killing most trees but melting a resin that allows the cone scales of the conifers to open. A new generation of seedlings soon dominates the site, and a forest stand will be reestablished often within a decade. Chaparral is another vegetation type that can be completely consumed by fire, but for some chaparral species, underground roots remain alive and new sprouts revegetate the site within half a dozen years. Other natural communities, such as lower montane conifer forests or coastal redwood forests, have an architecture that directs fire down to the ground, where low fires consume litter, herbs, shrubs, and saplings, burning relatively slowly and coolly. Roots, buried seeds, and burrowing animals are well insulated by the soil and survive.

Native Americans understood the role of natural fire, and they practiced sophisticated vegetation management of grasslands, woodland, and forests. This disappeared when Native Americans lost their land, along with their knowledge of fire-vegetation relationships. In the late 1800s, Euro-American loggers, miners, and sheepherders torched standing timber or piles of slash on windy summer days, creating disastrously intense fires that swept through the crowns of acres of forests, as natural ground fires would never do. As a result, the public came to view all fire as destructive, and new laws produced a century of fire-suppression. Combustible shrubbery has now risen so high beneath the canopies of many of our forests and in overgrown shrublands that ground fires quickly become all-encompassing. State and federal agencies are now working together to maintain natural ecosystems while protecting lives and property.

Michael Barbour and Valerie Whitworth

The vigor and abundance of death camas (Zigadenus fremontii) *can be startling amid charred shrubs.*

In spring, charred and blackened hillsides, TOP, *come alive with wildflowers that have been lying dormant as seed until the effects of a fire stimulate germination and bloom.* • *Phacelia* (Phacelia parryi), ABOVE LEFT, *and fire poppy* (Papaver californicum), ABOVE RIGHT, *are called fire followers as they are seldom seen except in areas swept by fire.* • *Beargrass* (Xerophyllum tenax), ABOVE MIDDLE, *rarely blooms except after a fire. It is shown here one year after a fire.*

WHY SO MANY RARE PLANTS?

Many of the plants that migrated to California died out or slowly emigrated when conditions changed. Others have persisted much as they were, while still others evolved into varied forms often restricted to areas with special climate, soil types, or elevations. These habitats often occur as isolated ecological islands surrounded by environments with different qualities. The presence of so many of these ecological islands, such as a cool coastal belt, a serpentine band in the Sierra foothills, or an isolated mountain top, has made California home to a large number of endemics, which is the name we give to species known only from a particular area. In fact, more than one-third of the state's native plants are found only in California. These species are considered endemic to California. If we consider the California Floristic Province a region that includes southwestern Oregon and northern Baja California but excludes the deserts and Great Basin, then the degree of endemism (forty percent) is even more remarkable for a continental flora. Alaska has only one endemic plant species, and New England has but a handful.

California's rugged topography and diverse climates have promoted floristic richness. Plant groups have evolved to adapt to the environmental conditions associated with 1,000 miles of sea coast, coastal mountains, hot dry valleys, the mighty Sierra Nevada, and parched deserts. A landscape fractured by topography and climate presents many barriers to plant migration, isolating populations from their close relatives, which in time can lead to the evolution of new species. As the Sierra Nevada uplifted millions of years ago, it created habitats that were colonized by plants that migrated westward from the Rocky Mountain region and southward from Canada, as well as by those that evolved from ancestors at adjacent lower altitudes. Intervening valleys isolated newly evolved endemics on high peaks; for example, each of the tufted rock cresses (*Arabis*) in the mustard family occupies its own small area of the Sierran crest.

Some of our most famous rare or restricted species, such as giant sequoia (*Sequoiadendron giganteum)*, Santa Lucia fir (*Abies bracteata)*, tree-anemone (*Carpenteria californica*), island ironwood (*Lyonothamnus floribundus*), Monterey pine (*Pinus radiata*), and Torrey pine (*P. torreyana*) were once more broadly distributed, but have retreated to their current ranges in response to climatic changes. These remnants of a past floral assemblage are called *paleoendemics*.

In contrast, many of California's endemics are rare because they have evolved only recently, often in geologically young or specialized environments of limited extent. These are called *neoendemics*. Meadowfoam (*Limnanthes*), for example, occurs in vernal pools that are probably only a few million years old at most. The Eureka Dunes evening primrose (*Oenothera californica* ssp. *eurekensis*), growing in the Mojave Desert, was only recently isolated from close relatives and has subsequently evolved into a distinct species.

Still other neoendemic groups multiplied in size and diversity in response to the availability of pollinators. Much of the diversity in the Polemoniaceae, a family containing phloxes and gilias, is attributed to the intertwined evolution of insects and plants.

About four million years ago, hot dry summers became the norm in California, and annual species experienced an evolutionary explosion. These plants flourish during favorable seasons and persist as seeds during seasons that are too cold, too hot, or too dry. They are especially suited to our extreme climate, and the hotter it gets, the more annuals are to be found: the flora of Death Valley, for instance, is nearly forty percent annuals. These newly evolved annuals, neoendemics, include seventeen rare and endangered phacelias and twenty-one monkeyflowers (*Mimulus*). One annual neoendemic, Merced clarkia (*Clarkia lingulata*), which is restricted to a small area in the Merced River Canyon, most likely evolved over the past 10,000 years.

California's floristic wealth is marked by more than 200 endemic plants known from fewer than five populations. Their scarcity and restricted range pose grave challenges for conservationists charged with maintaining the diversity of plant life in California. To name but a few, Chinese Camp brodiaea (*Brodiaea pallida*), Tiburon mariposa lily (*Calochortus tiburonensis*), and Mt. Diablo bird's-beak (*Cordylanthus nidularius*) grow only in one known location.

Beyond individual rare species, there are thirty entire genera of restricted or endangered endemic species found in the California Floristic Province and nowhere else. An entire subtribe of grasses, the Orcuttieae, with eight endangered species, occurs only in deep vernal pools in the California Floristic Province. The provincialism of the California flora, which thrills botanists and naturalists from around the world, is a blessing that carries with it a responsibility for us all.

Mark Skinner and G. Ledyard Stebbins

Red ribbons (Clarkia concinna), OPPOSITE, TOP, *in the evening primrose family is in another rapidly evolving group. • The beautiful snowdrop bush (Styrax officinalis),* OPPOSITE, MIDDLE, *named by the Greeks for its source of gum storax, has a very limited distribution in California. • The true lilies in California, derived from a single butterfly-pollinated ancestor, evolved into thirteen species with distinctive floral forms and are now pollinated by sphinx moths, butterflies, hummingbirds, and bees. Lilium parryi with the sphinx moth, Hyles lineata,* OPPOSITE, BOTTOM, *is shown here. • Once more widely distributed, western leatherwood (Dirca occidentalis),* TOP, *is now one of the rarest plants in California. • Bird's eyes (Gilia tricolor),* ABOVE, MIDDLE, *is in a genus with ninety-one species and subspecies in California, a phenomenon resulting from the closely intertwined evolution of plants and insects. • Colors of the bush monkeyflower (Mimulus aurantiacus),* ABOVE, BOTTOM, *range from deep reds and oranges to pure white. • There are 123 species and subspecies in the rapidly evolving genus, Phacelia, most of which have no common names. Phacelia minor, shown here,* LEFT, *grows in many places.*

PLANTS UNDER SIEGE: HABITAT LOSS BY MAN'S ACTIVITIES

"*T*he Great Central Plain of California . . . was one smooth, continuous bed of honey bloom, so marvelously rich that, in walking from one end of it to the other . . . your foot would press about a hundred flowers at every step. Mints, gilias, nemophilas, castillejas, and innumerable compositae were so crowded together that, had ninety-nine percent of them been taken away, the plain would still have seemed to any but Californians extravagantly flowery." So wrote John Muir as he set out to walk to the Sierra Nevada from his home in Martinez more than 100 years ago, crossing the Great Central Valley along the way.

Imagine his disappointment were he to walk this same path today. His journey would take him through blue and live oak woodland with an understory primarily composed of coarse weeds and non-native annual grasses, over hundreds of barbed wire fences separating one grazed pasture from the next, and across intensively farmed fields of tomatoes, wheat, safflower, and fruits and nuts. He would cross irrigation canals larger than many rivers, but might not find a single extensive patch of purple needlegrass (*Nassella pulchra*), once dominant over much of the northern Great Valley. In many parts of the valley's fertile portions, he would be hard pressed to find a single native herb in the scant hedgerows separating agricultural fields and newly built shopping centers and housing tracts.

Habitat loss and degradation have driven more than thirty plant species in California to extinction since John Muir's time. And although extinction is a natural process, the current rate of species loss, rather than the process itself, is cause for preeminent concern. Around the world, plant and animal extinctions have risen precipi-

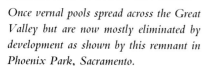

Once vernal pools spread across the Great Valley but are now mostly eliminated by development as shown by this remnant in Phoenix Park, Sacramento.

RARE AND ENDANGERED PLANTS IN THE CALIFORNIA FLORISTIC COMMUNITY

CNPS CATEGORIES OF ENDANGERMENT	TAXA	PERCENTAGE OF CA FLORA
1A. Presumed Extinct in California	34	0.5%
1B. Rare or Endangered in California and Elsewhere	857	13.6%
2. Rare or Endangered in California, More Common Elsewhere	272	4.3%
3. Need More Information	47	0.8%
4. Plants of Limited Distribution	532	8.4%
TOTAL	**1742**	**27.6%**

Source: CNPS *Inventory of Rare and Endangered Vascular Plants of California* (1994). Recent research for the 6th edition has identified nearly 600 additional potential candidates for inclusion in this table.

tously in recent times, and nearly all of these are attributable either directly or indirectly to human activities. Estimates vary, but human activities around the globe may well result in the loss of twenty percent of all species by the year 2050 in addition to drastic reductions in many species' ranges. This is much higher than the background rate of natural extinction, which would be far less than one percent during this same time span. And the decline and extinction of plant life are not restricted to tropical rainforests that teem with exotic and fantastic life forms, but are occurring at an accelerating pace in our own state.

Officially, the State of California lists only 216 plants as rare, threatened, or endangered under the California Endangered Species Act. Despite state protection, seventy-five percent of these listed plants continue to decline. Habitat loss is the single largest contributing factor.

As we enter the new millennium, nearly thirty percent of California's 6,300 native plants (1,742 species, subspecies, and varieties in all) are rare, threatened, or limited in distribution. More than 850 of these (fourteen percent of our native species) are considered rare or endangered by the California Native Plant Society.

Handsome valley oak (Quercus lobata), shown here being squeezed out by Christmas trees, were once common in deep valley soils but are becoming rare.

FREQUENCY OF NEGATIVE IMPACTS TO CALIFORNIA'S STATE-LISTED PLANTS

TYPE OF THREAT[1]	% OF STATE-LISTED PLANTS AFFECTED BY THREAT
Development (residential, industrial, commercial)	50.2
Livestock grazing	33.8
Off-road vehicles	25.8
Roads (construction and maintenance)	24.4
Agriculture	20.2
Exotic plants	19.2
Water projects	17.4
Fire management	16.4
Mining (sand, gravel, clay, minerals)	12.7
Trampling (by humans and horses)	11.3
Feral animals (pigs, goats, and introduced deer and elk)	6.6
Climatic effects, i.e., extreme weather events	6.6
Landfills (and trash-dumping)	5.6

[1] Other threats to state-listed plants cited at lesser frequencies include flood control activities, water quality degradation, logging, disease, horticultural collecting, energy development, hybridization, and vandalism.

Source: Department of Fish and Game's "Annual Report on the Status of California State-Listed Threatened and Endangered Animals and Plants" (1992).

URBANIZATION

Land development has fueled the surge of California's population and its economy; perhaps more than any other human activity, it has depleted local floras. The San Francisco Bay Area, the first region of California to be developed on a grand scale, is a haven for serpentine grasslands that support a number of endemic plants found nowhere else. These landscapes, dotted by tufts of purple needlegrass, are famous worldwide as wildflower gardens in the spring decorated by sheets of goldfields (*Lasthenia*), tidy-tips (*Layia*), California poppies (*Eschscholzia californica*), and fragrant lilies such as fritillaries (*Fritillaria*), brodiaeas (*Brodiaea*), and wild onions (*Allium*). Aerial photographs show that about half of this habitat has been converted to housing, industry, and roads. Along with the federally protected bay checkerspot butterfly (*Euphydrias editha bayensis*) and San Francisco garter snake (*Thamnophis sirtalis tetrataenia*), nearly forty highly restricted or rare plants grow on Bay Area serpentine soils and are vulnerable with the loss of this habitat. Fifteen of these are sufficiently threatened to be listed as rare or endangered by the State of California or the federal government.

Acres of houses now replace acres of oak woodlands in the Santa Clara Valley.

For every major metropolitan area in California, the same sad tale can be told: dwindling or degraded natural vegetation and consequent vulnerability of endemic plants. For instance, vernal pools on unique raised marine terraces in San Diego County have been largely destroyed; at most, less than ten percent remain. Many of those remaining are badly degraded by weeds, or by tire tracks that disrupt the delicate hydrological balance. Twenty-two rare plants occur only in this habitat, and nine are state or federally protected.

To the north, the lowland arid grasslands and scrublands of the Los Angeles basin are largely urbanized. Few undisturbed natural areas remain, and formerly common plants can be found only in small isolated patches. This vast area is home to ninety restricted or rare species, of which twenty-one are listed as endangered. Six are already extinct, including the San Fernando Valley spineflower (*Chorizanthe parryi* var. *fernandina*), which disappeared along with its coastal scrub habitat.

Coastal salt marshes once occurred in unbroken bands in protected estuarine locations from San Diego nearly to the Oregon border, but have since been filled to support our growing cities and industry. One fleshy salt-lover, the California sea-blite (*Suaeda californica*), formerly grew in the southern San Francisco Bay but was extirpated there by diking, draining, and building. It survives only in a small portion of Morro Bay in San Luis Obispo County. Point Reyes bird's-beak (*Cordylanthus maritimus* ssp. *palustris*), a partial parasite on asters and other shrubs, was once common in the salt marshes of San Francisco and Humboldt bays. With the loss of perhaps ninety percent of its habitat, it too has become scarce. In all, seventeen rare and endangered plants inhabit coastal salt marshes in California.

AGRICULTURAL PRODUCTIVITY

Our agricultural cornucopia feeds the nation: fine wines of the Sonoma and Napa region; San Joaquin Valley cotton and oranges; Sacramento Valley wheat, rice, safflower, and tomatoes; Imperial Valley melons; and Salinas Valley lettuce and strawberries. But the cost of this productivity, this conversion to agriculture, has been the near elimination of native plants, animals, and natural processes in the state's fertile bottomlands.

The native grasslands that once blanketed the Central Valley are ninety-nine percent gone, reduced to small preserves at places such as the Jepson Prairie in Solano County. Standing wetlands are ninety-five percent gone, having been filled and diked to support crops and orchards. California jewelflower (*Caulanthus californicus*) was once common throughout the San Joaquin Valley, but today it is an endangered species. So is Bakersfield cactus (*Opuntia basilaris* var. *treleasei*), driven nearly to extinction by agricultural production, energy development, and unchecked urbanization in the Bakersfield metropolitan area. The script is the same to the north, where vernal pool wetlands already severely degraded by agriculture are increasingly at risk as the greater Sacramento metropolitan area expands toward the Sierra Nevada foothills.

The San Joaquin Valley's agricultural productivity is world renowned but the cost to native plants has been devastating.

LIVESTOCK GRAZING

Inland, a debate rages. Is the livestock grazing that occurs across forty percent of California, or nearly forty million acres, a threat to California's native flora and rare plants? The paradoxical answer is both no and yes. The story begins long ago with the sixteenth-century Spanish who first brought cattle here, and also introduced to California, in animals' fur and in bales of hay, the first of more than 600 weedy species of Mediterranean origin. Gradually these Mediterranean exotics, now more than sixty percent of our weeds, came to dominate most grassy habitats in the state, driving out native annual and perennial grasses and herbs in the process. Only high-altitude meadows, inhospitable soils such as serpentine, and vernal pool bottoms— "too dry for aquatics, too wet for exotics"—are generally free from the scourge of weedy annual grasses. These grasses sprout early, grow tall, and rob native plants of moisture, sun, and space. But intermediate levels of

Sheep graze on the slopes of the Sierra Nevada, denuding former meadows.

grazing by cattle or native ungulates such as elk can shift the competitive balance and allow natives to thrive again. In the absence of grazing or fire, another natural disturbance to which native grassland plants are accustomed, weeds proliferate. Although cattle were originally involved in converting the grassland flora to an undesirable mix dominated by weeds, and heavy overgrazing continues to be detrimental to native species, moderate grazing now often sustains native grassland spring wildflower displays.

Altogether, more than 200 grassland, woodland, and meadow plants in California are declining because of the combination of heavy grazing and competition from Mediterranean exotics; seventy-four are state or federally protected.

On the summits of the Sierra Nevada and in highland areas of the state where most protected national forests and parks are situated, the chain of life continues largely unbroken for plants and their environments. But the valleys and coastal areas, where most land is privately owned, are more vulnerable to development and other activities destructive to natural habitat.

WATER DIVERSION

Nowhere is the contrast between highlands and lowlands more visible than east of the Sierra. Mount Whitney, protected in Sequoia-Kings Canyon National Park, looms over the Owens Valley two miles below, where water diversion by the City of Los Angeles and heavy grazing have caused dust storms and drastically reduced suitable wet habitat for the endangered Inyo County star-tulip (*Calochortus excavatus*), Fish Slough milk-vetch (*Astragalus lentiginosus* ssp. *piscinensis*), and Owens Valley checkerbloom (*Sidalcea covillei*).

Today in California we confront a natural world where once common species are now rare, and rare species are now endangered. We risk modifying our landscape to such a degree that the ability of native organisms to survive, and ultimately to evolve, may be fatally compromised. Yet proper management of exotic plants and animals can be achieved if we wish to spend the money. Large-scale efforts to restrict development to less environmentally sensitive areas, to establish large protected areas in diverse habitats, and to preserve corridors for the dispersal and migration of plants and animals should be intensified. Both Federal Habitat Conservation Planning and California's Natural Communities Conservation Planning are thoughtful approaches that may secure viable habitat in rapidly deteriorating parts of California, such as the Owens Valley, the Central Valley, and coastal Southern California. While harmony with nature remains possible, too many species are becoming scarcer day by day.

Mark Skinner and Peggy Fiedler

CALIFORNIA'S LIVING LANDSCAPE: VEGETATION TYPES AND PLANT COMMUNITIES

The extraordinary botanic wealth of California is a reflection of environmental richness unparalleled anywhere in the temperate world. With both the lowest and the highest points in the continental United States, spanning more than ten degrees of latitude with over 1,100 miles of coastline, and covering over 100 million acres, California today is a bridge between foggy, dimly lit rainforests and open, parched, sunbathed deserts. The interaction of climate, soil, bedrock, and elevation creates a mosaic of vegetation types that characterize California's regional landscapes.

Vegetation is the plant cover of a region, a thin clothing over the land that is at once fragile and durable, able to repair and reproduce itself over centuries. In contrast to many other states, California has dozens of vegetation types. Each is named after a geographic location and a dominant plant form—for example, riparian forest, upper montane conifer forest, Central Valley annual grassland, foothill woodland, alpine tundra, coastal salt marsh, and warm desert scrub.

A lingering relic of the past, the Torrey pine (Pinus torreyana) *grows only along a narrow coastal strip in San Diego County and on Santa Rosa Island.*

Each type also has a characteristic architecture created by the growth forms of the dominant plants and the layering of associated species. For example, lower montane conifer forest grows at middle elevations throughout the mountains of California. This vegetation type has four layers of plant canopies: a patchy overstory of large trees such as ponderosa pine (*Pinus ponderosa*), Douglas-fir (*Pseudotsuga menziesii*), incense cedar (*Calocedrus decurrens*), sugar pine (*Pinus lambertiana*), and white fir (*Abies concolor*) about 150 feet tall; a scattered understory of winter-deciduous black oak (*Quercus kelloggii*) and mountain dogwood (*Cornus nuttallii*) twenty to thirty feet tall; a denser shrub layer with greenleaf manzanita (*Arctostaphylos patula*) and dwarf tanbark oak (*Lithocarpus densiflorus* var. *echinoides*); and an herbaceous layer of perennials such as violets (*Viola* spp.) and coralroots (*Corallorhiza* spp.). Although each canopy is incompletely closed, most of the ground is obscured from above by the combination of layers.

Within a vegetation type there usually are smaller units called plant communities. Each plant community is characterized by several species that make up the dominant plant form, and the community is named for these species. Communities within the lower montane conifer forest include ponderosa pine forest, Douglas-fir forest, white fir forest, mixed conifer forest, and giant sequoia forest. Wherever a particular habitat repeats itself, the same general cluster of species recur, generally with the same architecture.

Valley oak (Quercus lobata) *is the monarch of all deciduous oaks; it once clothed the valleys of California from Mendocino south to San Fernando and Pasadena and particularly the wide Central Valley where deep loam and high water tables favored their growth into magnificent spreading shapes.*

Every community, then, has a predictable habitat, species composition, and vegetation structure. Some communities have such narrow habitat requirements that they are rare, such as serpentine grassland, while others are widespread and common, such as Douglas-fir forest or oak woodland.

According to a plant community classification used by the California Native Plant Society and the California Department of Fish and Game, more than 300 plant communities are found in California. Many of these attract visitors to California from all over the world: dripping-wet coast redwood (*Sequoia sempervirens*) forest along the north coast; twisted Monterey cypress (*Cupressus macrocarpa*) forest on the Monterey peninsula; Torrey pine (*Pinus torreyana*) groves at the lip of ocean-facing bluffs; oak woodlands in foothills of Coast and Peninsular ranges; vernal pools that add brilliant color to spring grasslands; looming giant sequoia (*Sequoiadendron giganteum*) forests of the Sierra Nevada; bristlecone pine (*P. longaeva*) subalpine woodland high in the White Mountains; an ocean of sagebrush (*Artemisia tridentata*) scrub throughout northeastern California; and fan palm (*Washingtonia filifera*) oases in remote desert canyons.

PLANT COMMUNITIES AND SPECIES DIVERSITY

Plants occur where they do not by chance, but because the environment meets their requirements. A nearby area that appears similar yet does not support a particular species, upon closer examination often differs in subtle but important ways: soil depth, texture, or nutrient status; history of disturbance; slope direction; intensity of shade; or availability of moisture. Floral rarity can thus result from a particular set of physical conditions.

Diversity of habitats or plant communities is quite different from diversity at the species level. A community can be rare without containing a rare plant species if the particular mix of species is rare. If we focus only on rare species, we might miss a rare habitat, a rare assemblage, or a habitat critical to a rare species.

In addition, the ecological importance of one species often extends through its community and is thus magnified. Neighboring plants that have symbiotic or competitive relationships with a species will be affected by what happens to that species, as will insects that pollinate its flowers, soil microbes that digest its litter, or grazing animals that feed on its roots or stems. It is best to assume that every species within a community plays some role in the existence of others in that community.

Sixty million years ago, when California was less mountainous and the climate more tropical, flannelbush (Fremontodendron californicum) was more widespread.

CALIFORNIA'S VEGETATION: LEGACY OF CHANGE

California's landscapes have changed dramatically over the vast panorama of geologic time and the vegetation has reflected these changes. Topography has been modified by fluctuations in sea level, periods of mountain building and faulting, glacial advances and retreats, and the movement of the continents.

Sixty million years ago the region was less mountainous and the climate was more tropical. Tree ferns, palms, cycads, and large-leaved tropical plants —plants whose closest relatives today occur in southern Mexico and Central America—dominate the fossil record from that period. Only a few relicts from the fossil record still remain in California, for example, fan palms and ironwood trees (*Olneya tesota*) in the desert and flannelbush (*Fremontodendron californicum*) in the foothills.

Forty million years ago the topography was more varied and the climate was cooler than today. Fossils of spruce, pine, fir, and winter-deciduous hardwoods such as beech, elm, and maple dominate the geologic record from that time, indicating that rich, mixed conifer-hardwood forests covered the land. Relatives of these fossil plants still occur in California along the cool, wet coast and at high elevations in the mountains. Some famous endemic California species, such as coast redwood (*Sequoia sempervirens*) and giant sequoia (*Sequoiadendron giganteum*), date from this time.

About ten million years ago a semi-arid Madro-Tertiary geoflora became dominant in the fossil record, including close relatives of the modern madrone, live oak, pinyon, and chaparral and desert shrubs, which apparently moved into California from Mexico as our Mediterranean climate became more pronounced. Today descendants of this flora dominate low-elevation vegetation throughout much of the state. Most of our endemic plants evolved in place over the past several million years from this semi-arid flora.

One to two million years ago an ice age, with several glacial advances and retreats, produced cooler temperatures than at present, and desert vegetation was pushed far to the south. Many montane features were carved then by glaciers that descended as low as 4,000 feet. The last glacial retreat ended 10,000 to 12,000 years ago, coinciding with the arrival of human immigrants from Asia and the disappearance of many large grazing animals. About 3,000 to 8,000 years ago the climate became drier and several degrees warmer than at present. As a result, desert vegetation expanded and timberlines were pushed to higher elevations.

Precipitation and temperatures have continued to fluctuate in the past few thousand years, but within narrower limits. According to one study of tree rings from big-cone Douglas-fir in Santa Barbara County, there has been no cumulative, long-term change in rainfall in California over the last 400 years. There have been decades of wet or dry weather. A drought during the period 1840-60 was especially damaging because it coincided with the rapid influx of Euro-Americans, the conversion of wildlands to agriculture, the introduction of aggressive weedy plants, the replacement of native grazing animals with domesticated livestock and an increase in grazing intensity, and a change in the frequency and intensity of fires. Many landscapes were permanently damaged during this time, and degradation has continued to the present. Current evidence for a global warming trend with increases in the concentration of carbon dioxide by nineteen percent over the past 100 years, possibly resulting from human activity, makes speculation about California's native plant species and vegetation even more complicated.

In recent history, explorers and colonists either accidentally or deliberately brought with them plants from their countries of origin. Approximately 1,000 of these non-native species now grow naturally in California, some in such high numbers that they have changed forever the face of the landscape. Grasslands, for example, were once dominated by native perennial bunchgrasses, but they now consist largely of introduced weedy annual grasses and other types of plants.

There is hardly a plant community or habitat left in California unaffected by some introduced plant species. European beachgrass (*Ammophila arenaria*) smothers coastal dunes; Eurasian salt cedar (*Tamarix* spp.) chokes riparian areas; broom (*Genista* spp.), gorse (*Ulex europaea*), and Chinese tree-of-heaven (*Ailanthus altissima*) all crowd into forest edges; pockets of Australian eucalyptus spread into foothill woodland; and annual cheat grass (*Bromus tectorum*) from the steppes of Asia now carpets the sagebrush desert of northeastern California.

John Fremont's accounts of his explorations of nineteenth-century California are replete with references to colorful vegetation found along riparian corridors in fall.

INVASIVE WEEDS: AN INSIDIOUS THREAT TO CALIFORNIA WILDLANDS

California's most beautiful wildlands are under siege by aggressive biological invaders. Introduced weeds, also known as exotic or alien pest plants, have invaded millions of acres throughout California, threatening not just agricultural lands, but natural habitats and rare plants as well. European beachgrass, South American pampas grass, and South African iceplant are taking over coastal dunes; French broom has invaded North Coast shrublands and woodlands; tamarisk from the Middle East pulls precious water from desert springs; oceans of cheat grass increase fire frequency in sagebrush east of the Sierra Nevada and giant reed grass has crowded out native riparian plants in many California streams. Many of the habitats being overrun by these exotic invaders are home to rare and endangered plants and wildlife. Weeds pose a direct threat to at least forty-nine of California's 216 listed rare, threatened, and endangered plants.

Exotic pest plants in California have caused large-scale ecosystem changes, including altered fire and water cycles. Weedy eucalyptus trees, brought here from Australia, became naturalized and have invaded coastal stream habitats where they shed enormous quantities of flammable bark, shade out and chemically inhibit the establishment of other plants, and by consuming more water than drought-tolerant natives, draw down the water table. The last populations of the rare marsh sandwort (*Arenaria paludicola*) are threatened by stands of weedy eucalyptus; local efforts have begun a tree removal program to restore the marsh sandwort's last retreat.

Perennial pepperweed (*Lepidium latifolium*), also known as tall whitetop, is a Eurasian native that has recently and rapidly invaded coastal and interior marshes. Its abundant seeds germinate readily and it rapidly forms a massive taproot. Although it cannot invade a fully tidal salt marsh, stands of tall whitetop border the "high marsh" habitat that is home to the rare soft bird's-beak (*Cordylanthus mollis* ssp. *mollis*), a plant that has mysteriously declined in recent years. Researchers have speculated that gradual, long-term changes in freshwater inflow and elimination of tidal action are providing conditions that can be exploited by tall whitetop at the expense of less aggressive marsh plants such as soft bird's-beak.

Disturbance has allowed many weedy invaders to gain a foothold on wildland borders, but the worst culprits are able to spread into more pristine areas where recent disturbance is not apparent. French broom and its Mediterranean relatives, Scotch broom and Spanish broom, with their masses of golden yellow flowers, were first brought to Northern California after the gold rush to be used in landscaping. Well adapted to California's climate, these prolific

Garden escapes such as this calla lily (Zantedeschia aethiopica) *sometimes naturalize and become aggressive weeds, excluding native plants and altering native plant communities.*

seeders rapidly escaped from gardens to roadsides; some were even planted on roadcuts until 1978. Today brooms have invaded conifer and oak woodlands and have replaced native shrubs in our coastal scrub communities. Where dense stands of broom eliminate the native flora, they create a serious fire hazard.

Tamarisk is another Eurasian weed that degrades wildlands. It infests over a million and a half acres of wetlands and stream courses in the Southwest, including California's deserts. Tamarisk is a water waster, drying up desert springs in such sites as the Dos Palmas Preserve in Riverside County. Aggressive removal efforts have restored the Dos Palmas springs that harbor endangered desert pupfish.

Have you seen a redwood forest with a pampas grass understory? The silvery plumes of this giant South American grass now wave within most of our coastal communities: dunes, forests, and shrublands alike. Imported to California around the turn of the century to provide decorative plumes for Victorian parlors, pampas grass spreads by millions of tiny windblown seeds. It threatens many of our endemic dune plants—fragile and low-growing wildflowers such as surf thistle (*Cirsium rhothophilum*), sand gilia (*Gilia tenuiflora* var. *arenaria*), and beach spectacle-pod (*Dithyrea maritima*)—which are overwhelmed by the spreading clumps of this gargantuan grass.

Concerned citizens have fought back against pampas grass and other botanical barbarians. Pampas grass has so changed the landscape of coastal counties that many have organized "grass-roots" opposition, hacking and spraying to rid their communities of this invader. On the North Coast landowners are offered environmentally benign replacement plants if they allow eradication crews to remove pampas grass from their properties.

In other parts of the state, local groups are working to control or eradicate tamarisk, giant reed grass, broom, iceplant, and other plant pests. Habitat restoration is the ultimate goal of these efforts, in order to preserve the biological balance and natural diversity at risk from the global pollution of wildlands by exotic pest plants.

Ann Howald

Some of the most aggressive weeds in California include pampas grass (Cortaderia jubata), TOP, *South African iceplant* (Carpobrotus edulis), MIDDLE, *and gorse* (Ulex europaeus), BOTTOM, *in the coastal dune and scrub communities. Miles of pampas grass crowd out native plants along Highway 1 along the central coast, and the rare Howell's spineflower* (Chorizanthe howellii), MIDDLE, *is being crowded out by a carpet of South African iceplant, and pastures are made unusable by gorse.*

COASTAL MOSAIC OF DUNES, WETLANDS, PRAIRIE, AND FORESTS

C alifornia's western edge is a restless interface between land and sea. Landforms along the 1,000-mile coastline are spectacular and diverse: sandy bays, muddy tidal flats, precipitous cliffs, undulating dunes, terraced grasslands, steep shrub-covered hillsides, and dense forests fringing riverine bottomlands. The coastal mosaic of vegetation covers about fifteen percent of the area of California.

BEACHES AND DUNES

Beach plants at the leading edge of vegetation must be tough enough to withstand the volatile coastal environment. Salt spray, abrasive sand blast, sand substrate, low soil nitrogen, and high light intensity are some of the stresses strand plants face. As a result, beach vegetation is open and prostrate, made up of a handful of species at any one location. Gentle sand hummocks are splattered with perennial herbs that spread vegetatively by runners and rhizomes or by clumps of beach grass. Most of the grass is European beachgrass (*Ammophila arenaria*), brought here in 1869 to stabilize sand dunes for San Francisco's Golden Gate Park, and now the most abundant beach and dune plant from Big Sur to Oregon.

Beach sagewort (Artemisia pycnocephala), *a white-woolly, native plant (shown in foreground), commonly colonizes and stabilizes dunes from Monterey County north.*

WETLANDS

Prior to 1900 there were four to five million acres of wetlands in California. Today a fragmented less than ten percent of our coastal wetlands and a meager two percent of our interior wetlands still exist. Of these, ten percent are severely degraded. It is no coincidence that twenty-five percent of plant species and fifty-five percent of animal species designated by state agencies as threatened or endangered have wetlands as their essential habitat. The term wetland does not mean that a single vegetation type or a single plant community will occur there. As a legal term it currently applies to any habitat where the soil surface is saturated with water to within eighteen inches of the surface for a period of at least one week a year. Wetlands are periodically waterlogged, and plants that grow there must be tolerant of low levels of soil oxygen.

The tidal salt marsh found at China Camp State Park in Marin County is one of the most pristine in the state.

Only a small subset of California's flora is flood-tolerant. The presence of flood-tolerant species is a good indication that the site is a wetland even if the ground appears to be dry most of the year. Other kinds of wetlands include seagrass beds, tidal salt marshes, brackish water marshes, freshwater tule marshes, vernal pools, riparian forests, montane wet meadows, and desert oases and playas.

COASTAL PRAIRIE AND SCRUB

Inland from the beaches California's coastal cliffs, terraces, and rolling hills are covered with different species, growth forms, and plant communities. This is the region of coastal prairie and coastal scrub.

Spring brings a medley of wildflowers, owl's-clover, tidy-tips, and bush lupine, to soft, chaparral-covered coastal terraces and prairies.

In the mid-nineteenth century, with the arrival of settlers, the coastal prairie was transformed into a grassland of introduced species. Euro-Americans brought fire suppression, invasive weeds, and heavy year-round grazing by cattle and sheep. Weedy herbs, particularly those that cattle find unpalatable, spread indiscriminately in the absence of fire. In this way milk thistle, wild artichoke, and Klamath weed came to dominate millions of acres of former coastal prairie, drastically reducing the biological and economic value of the land. Shrub-dominated vegetation or scrub covers steep slopes and may invade the margins of coastal prairie on terraces and in valleys. Northern coastal scrub extends north of Big Sur. It is a dense, two-storied assemblage of shrubs, vines, herbs, and grasses. Dominant plants are coyote brush (*Baccharis pilularis*), salal (*Gaultheria shallon*), California coffeeberry (*Rhamnus californica*), cow parsnip (*Heracleum lanatum*), and bush lupine (*Lupinus arboreus*). This vegetation also has been modified in the past century by introduced plants, especially pampas grass, gorse, Scotch broom, and eucalyptus.

Southern coastal scrub extends south of Big Sur. It is a lower, more open community than northern scrub, and it is dominated by drought-deciduous sage (*Salvia* spp.) and sagebrush (*Artemisia californica*) species. Near the Mexican border succulent plants are an added element: century plant (*Yucca* spp.), live-forever (*Dudleya* spp.), and several kinds of cactus. This vegetation type is fast disappearing, as the gentle hills and terraces of this community are being converted into housing and commercial developments.

COASTAL FOREST

A series of coastal forests, including redwood (*Sequoia sempervirens*), Douglas-fir (*Pseudotsuga menziesii*), and closed-cone conifers, occupy lowland pockets and mountain slopes farther from the ocean. These are unique forests, relics from ancient times, coddled by a moderating maritime environment in a sea of seasonal fog.

Conifer trees have the fastest growth rates of any trees in the world. The most famous conifer in California is the coast redwood, forests of which form the southern tip of a magnificent conifer forest that blankets a strip of coastal slopes and flats from Alaska to Monterey County.

Found growing in the fog belt along the coast, the coast redwood (Sequoia sempervirens) *is the tallest tree in the world.*

Old-growth redwood forests are uncommon today. Only five percent of the original two million acres remain, mostly in state and federal parks. The rest have been logged once or twice and support second-growth redwood forest or young forests converted to faster-growing Douglas-fir. Most of the old-growth Douglas-fir phase of the mixed evergreen forest has been cut for timber. The results are dense stands of pole-size hardwoods, especially tan oak (*Lithocarpus densiflorus*) and madrone (*Arbutus menziesii*), since these are capable of stump-sprouting following disturbance. Young second-growth forests are quite different ecologically from old-growth forests.

Closed-cone conifer stands are another type of forest occurring in coastal mountains. Most closed-cone conifers are coastal in distribution, and grow on droughty or nutritionally poor sites with sandy, salty, acidic, shallow, or serpentine soils. They are small trees with life spans of seventy-five to 300 years. A few have economic value; for example, Monterey pine (*Pinus radiata*) is planted as a lumber tree in many parts of the world, and Monterey cypress (*Cupressus macrocarpa*) is widely planted as an ornamental. Other closed-cone conifers include knobcone pine (*P. attenuata*), Coulter pine (*P. coulteri*), Sargent cypress (*C. sargentii*), and the pygmy cypress (*C. pygmaea*) of Mendocino County, which forms the overstory of a scraggly forest sometimes only three or four feet tall.

INTERIOR MOSAIC OF GRASSLAND, CHAPARRAL, AND WOODLAND

Inland from the fog belt in the Coast Ranges, summer heat becomes more extreme and a typical Mediterranean climate prevails. The texture of interior foothills and valleys is a patchwork: woodland with scattered trees of gray-green pine and blue-green oak, muddy green chaparral with densely rigid shrubs, or open and golden-brown grassland. Collectively, the three vegetation components of this interior mosaic cover one-third of California.

While all three vegetation types occur within the same climatic and elevation zones, they vary with soil depth, soil texture, and slope face. Fire history is another important factor in creating this landscape, and many species have adapted to fire. Grassland tends to occupy the gentlest slopes with the deepest and most finely textured soils, and it may burn every few years. Chaparral tends to cover the steepest slopes with the shallowest and coarsest soils, and it burns with an intermediate frequency of fifteen to twenty-five years. Foothill woodland grows on intermediate slopes and soils, and it burns at intervals longer than twenty-five years, perhaps longer than fifty years.

INTERIOR GRASSLAND

The original interior grassland blanketed much of the Great Valley and low elevations along the central and southern coast. It covered more than thirteen million acres and an additional nine to ten million acres underneath oaks in the foothills for a total area representing one-fourth of California. The interior grassland was dominated by several species of bunchgrasses, particularly purple and nodding needlegrasses (*Nasella* spp.). Between the grasses were some annual and perennial herbs, lush and colorful in spring.

This grassland no longer exists, except in very small preserves that show only a portion of the pristine panorama. Beginning in the nineteenth century, livestock were kept in large numbers, year-round, in fenced pastures. Some grassland was plowed and farmed. Fire was controlled, and weed seeds were accidentally introduced. In a short time, the bunchgrass prairie was converted to an annual grassland of introduced European plants.

VERNAL POOLS

Vernal pools are ecological islands in a sea of grass. Found on gently rolling topography underlain with a hardpan that impedes water percolation, they fill with water in winter, supporting a meadow vegetation that grows up through standing water to flower in rings around the pool as water evaporates in spring. Every Northern California county in the Central Valley contains vernal pools, as do several Southern California counties. These pools occur within other vegetation types, such as oak woodland and chaparral, as well as in grassland. Two hundred years ago, vernal pools may have covered one percent of California; although estimates vary, conversion to agriculture and urbanization have probably removed between seventy to ninety percent of this habitat.

Owl's-clover (Castilleja densiflora) *and goldfields* (Lasthenia californica) *abound on the remnant grasslands found on the Carrizo Plain Natural Area in San Luis Obispo County.*

In spring great washes of color brighten the grasslands beneath the oak-dominated foothill woodlands.

FOOTHILL WOODLAND

If Californians were to select a symbolic vegetation type, foothill woodland would be the logical choice because of its historic importance, the large area it covers, its widespread familiarity, and its endemic trees. This is a two-storied vegetation with an open tree canopy shading only a third of the ground and with a carpet of grasses and wildflowers beneath. Close to the coast the dominant trees are Oregon oak (*Quercus garryana*) and California black oak (*Q. kelloggii*) in the north, and coast live oak (*Q. agrifolia*), Engelmann oak (*Q. engelmannii*), and California walnut (*Juglans californica*) in the south. Along drier interior foothills, the dominant trees are blue oak (*Q. douglasii*), interior live oak (*Q. wislizenii*), valley oak (*Q. lobata*), gray pine (*Pinus sabiniana*), and buckeye (*Aesculus californica*). The trees are a mix of evergreens and deciduous hardwoods. The deciduous species, such as blue

27

oak and valley oak, seem to be in decline; few young trees have become established in the last century. If this pattern continues for another century, mature oaks will have reached the end of their natural life spans, and these oak woodlands will become either grasslands or more dense woodlands dominated by evergreen oaks. The causes of this decline, not yet well understood, may include exploding populations of root-eating pocket gophers resulting from lack of seasonal flooding, browsing deer and cattle, and non-native annual plants that compete for nutrients and water.

CHAPARRAL

Chaparral vegetation is a single layer of almost impenetrable shrubs four to eight feet tall with intricately branched, interlacing evergreen canopies. The most common shrubs in Northern California chaparral are chamise (*Adenostoma fasciculatum*), scrub oak (*Quercus dumosa*), Christmas berry (*Heteromeles arbutifolia*), California coffee-berry (*Rhamnus californica*), and more than twenty species each of manzanita (*Arctostaphylos* spp.) and ceanothus (*Ceanothus* spp.). Yucca (*Yucca* spp.), red shank (*Adenostoma sparsifolium*), laurel sumac (*Malosma laurina*), and lemonadeberry (*Rhus integrifolia*) are additional elements in Southern California. The ground itself is nearly bare of plants, and only an occasional bay tree (*Umbellularia californica*), clump of cypress (*Cupressus* spp.), or knobcone pine (*Pinus attenuata*) overtops the dense shrub layer.

Seventy-five percent of California's watersheds are covered with shrubby chaparral vegetation, which significantly increases water retention in the watershed and reduces floods from fast water runoff.

MOSAIC OF MONTANE VEGETATION

California mountains are not uniformly covered with a continuous forest, but are a mosaic of oak-filled canyons, brushy ridges, meadows in wet flats, riparian woods, and conifer-forested slopes. Only about half of our montane area, which covers twenty percent of California, supports conifer forest.

Four zones of climate and vegetation are found in California's mountains, changing with increasing elevation: lower montane, upper montane, subalpine, and alpine. Every thousand-foot climb in elevation is equivalent to moving 300 miles north, and the vegetation varies accordingly.

LOWER MONTANE MIXED CONIFER FOREST

A mixed conifer forest occupies much of the lower montane zone that lies between 2,000 and 5,000 feet in northern mountain ranges and between 5,000 and 8,000 feet in southern ranges. Frost is common in this zone, and about one-third of the annual precipitation falls as snow. The growing season is six to eight months long. Six conifers—ponderosa pine (*Pinus ponderosa*), sugar pine (*P. lambertiana*), white fir (*Abies concolor*), Douglas-fir (*Pseudotsuga menziesii*), incense cedar (*Calocedrus decurrens*), and giant sequoia (*Sequoiadendron giganteum*)—coexist and shift in importance from stand to stand. For this reason, the community is not named for any one of them.

John Muir, among other early naturalists, described these forests: "The inviting openness where trees stand more or less apart in groves, or in small irregular groups, enabling one to find a way nearly everywhere, along sunny colonnades and through openings that have a smooth, park-like surface..."

The forests Muir saw are rare today because of logging and fire suppression. Half the original acreage of the mixed conifer forest has been cut at least once in the last 150 years. The overstory conifers once averaged more than three feet in diameter and 100 feet in height, but second-growth forests are more crowded, with smaller overstory trees.

UPPER MONTANE FOREST

The upper montane zone receives the maximum amount of snowfall in California. About eighty percent of annual precipitation falls as snow, building an eight- to thirteen-foot snowpack that stays on the ground up to 200 days. Saplings must tolerate burial until they are tall enough to stand above the snow.

Upper montane forests are simple, with just two canopies: an overstory tree layer and a scattered herb layer. Common trees are lodgepole pine (*Pinus contorta*) and Jeffrey pine (*P. jeffreyi*) throughout the state (5,500 to 7,500 feet elevation in the north, 8,000 to 10,000 feet elevation in the south), joined by red fir (*Abies magnifica*) and western white pine (*P. monticola*) in Northern California. Ancient, twisted Sierra juniper trees (*Juniperus occidentalis* var. *australis*) seem to spring full grown out of solid ridge rock. Quaking aspen (*Populus tremuloides*) and black cottonwood (*P. balsamifera* ssp. *trichocarpa*) cover narrow riparian corridors. Their shimmering, deciduous canopies turn golden in fall. Aspen can also invade clear-cut forests, where conifers once dominated, but conifer seedlings gradually appear in the aspens' shade and, over time, the site returns to conifer forest.

MIXED SUBALPINE WOODLAND

Trees reach their upper elevation limit in the subalpine zone (7,500 to 10,000 feet elevation), where only a few trees can tolerate the special stresses. Lodgepole pine and western white pine continue upslope from the upper montane forest, joined by mountain hemlock (*Tsuga mertensiana*), the five-needled whitebark pine (*Pinus albicaulis*), and limber pine (*P. flexilis*), foxtail pine (*P. balfouriana*), and bristlecone pine (*P. longaeva*), which are endemic to the subalpine zone.

The composition of mixed conifer forests varies with elevation and with other physical factors such as soil, moisture, and exposure.

Fall colors of the leaves of alders, aspens, and cottonwood brighten the creeks and streams in montane forests.

29

Occasional stands of mountain hemlock (Tsuga mertensiana) *can be recognized by their nodding tops and gracefully drooping branches.*

Not all of these trees are found together on the same mountain. Sometimes a single tree will dominate; elsewhere they mix in different ways. For this reason the vegetation is called mixed subalpine woodland. Of all subalpine species, only lodgepole pine is found from Oregon to Baja California. Subalpine tree seedlings struggle to survive because conditions are so stressful. Few succeed, but once established, they grow very slowly for centuries. Average life spans are 500 to 1,000 years. Foxtail pines, for example, attain ages greater than 3,000 years, and bristlecone pines of the White Mountains can exceed 5,000 years.

ALPINE TUNDRA

The alpine zone (10,300 to 11,000 feet elevation or more) is a thin fringe of green tundra near the limits of plant life. The growing season is measured in weeks and days, not months: six to ten weeks or forty to seventy days. Even during the growing season, frost can occur nightly. Plants hug the ground and sequester most of their stored food underground in roots or rhizomes.

Alpine tundra is a meadow-type of vegetation rich in sedges, rushes, grasses, and perennial herbs. Woody dwarf willows and white and mountain heather are present but not dominant. Away from wet basins and areas of snowmelt, the environment becomes drier and tundra gives way to a fell-field of scattered bunchgrasses and cushion plants—herbs with tiny, densely arranged leaves pressed against the soil surface, such as buckwheat (*Eriogonum* spp.), spreading phlox (*Phlox diffusa*), cushion cress (*Draba lemmonii*), and pussypaws (*Calyptridium umbellatum*). Alpine plants are small but tenacious, and they can live twenty to fifty years.

In California about forty percent of the total number of alpine species are widespread, ranging north into the Canadian Rockies and, in some cases, into the polar tundra. Another fifteen percent are endemic to California, and must have evolved as the mountains rose during the past ten million years. These endemics probably evolved from desert plants downslope.

Mat-forming Davidson's penstemon (Penstemon davidsonii) *grows in the talus on Mount Shasta with a low cushiony form suitable for harsh mountain weather.*

GRADIENTS OF DESERT VEGETATION

Desert-facing slopes of California mountains are steep, rocky, and more arid than other slopes at the same elevation. The eastern rainshadow intensifies as elevation drops, and by 6,000 feet trees reach a low-elevation timberline. This timberline, the last gaps of montane trees before the unrelenting aridity of desert scrub below, is an open woodland of Utah juniper (*Juniperus osteosperma*) and single-leaf (*Pinus monophylla*), two-leaf (*P. edulis*), and four-leaf (*P. quadrifolia*) pinyon pines.

Beyond the pinyon-juniper woodland is desert scrub vegetation, occupying about one-fourth of California's land area. Desert climate and vegetation are not uniform within such a vast region. There are three types of desert in California: cold, warm, and hot.

COLD DESERT

The cold desert, or Great Basin Desert, is a high desert lying above 4,000 feet elevation between the Sierra-Cascade axis on the west and the Wasatch Mountains of Utah on the east. Winters are cold, with most precipitation falling as snow. Summers are dry and warm. A land of dark volcanic rock, pronghorn antelope, bunchgrasses, and sagebrush, it is ecologically similar to the semi-arid steppes of central Asia. Cold-desert vegetation has been significantly modified over the past century by grazing, accidental introduction of aggressive weedy species such as cheat grass (*Bromus tectorum*), and a resultant increased frequency of wildfire.

Summers are hot and dry, winters are cold with some snow in the Great Basin Desert, a cold desert, which lies above 4,000 feet elevation.

WARM DESERT

The warm Mojave Desert lies just south of and at elevations below those of the cold Great Basin Desert. The meeting ground of the two is the place of Joshua tree woodland and blackbrush scrub but, as elevation drops further, the landscape becomes dominated by creosote bush (*Larrea tridentata*), burro bush (*Ambrosia dumosa*), brittlebush (*Encelia farinosa*), more than twenty species of cactus, and a great diversity of winter annual wildflowers. Localized habitats such as sand dunes, saline basins, and desert washes support rarer plant species.

The Mojave Desert comes alive in spring with the flowering shrub brittlebush (Encelia farinosa).

HOT DESERT

The hot Colorado Desert occupies the southeastern corner of California at elevations below 1,000 feet. The region receives summer rains from subtropical storms originating in the south, and the terrain is low enough to be frost-free. As a result, the Colorado Desert is home to a greater variety of plant growth forms and species than the other two deserts. Winter-deciduous small trees, cacti, evergreen and drought-deciduous shrubs, rosette-shaped succulents, subshrubs, small cacti, and winter and summer annual wildflowers all grow here. The overwhelmingly dominant species are creosote bush and burrobrush.

Springtime brings a breathtaking burst of yellows, pinks, reds, and all colors to the San Ysidro mountains in Anza-Borrego State Park located in the Colorado Desert.

WHAT LIES AHEAD?

From a distance, as from a jetcraft flying over the state at great elevation and speed, the changes that have occurred in California's vegetation since the first immigrants arrived two or three centuries ago may not be immediately evident. Areas of oak woodland, desert scrub, montane forest, chaparral, woodland, and grassland still exist. Urbanization, agriculture, clear cuts, and pavement have not replaced all our natural vegetation. Although only ten percent of old-growth forests, less than ten percent of coastal wetlands, and two percent of interior wetlands still exist, those vegetation types remaining are significantly different from their ancestors.

Desert scrub has been affected by grazing and off-road vehicles. Montane forests have flammable, dense understories because of fire suppression and overstories weakened from drought, insects, and atmospheric pollutants. Hundreds of square miles of chaparral have been converted to grassland or homesites, coastal scrub to suburbs, and perennial grasslands to weedy annual pastures or farmland. The quality of remaining fragments has been affected by nearby development.

It is time to question our relationship with the environment. Is it prudent to degrade whole ecosystems in our rush to settle the land, or to sacrifice long-term ecosystem stability for short-term benefit? To reclaim some portion of the original natural richness of California we need to restore and enhance areas beyond those that exist in preserves today.

Water buttercup (Ranunculus aquatilis) *adds beauty to a pond at Lassen National Park.*

Michael Barbour and Valerie Whitworth

AT A CROSSROADS: CAN CALIFORNIA'S NATIVE PLANTS BE CONSERVED?

Californians face the daunting challenge of protecting one of the world's most diverse and unique floras amid the pressures imposed by the nation's most populous state. Over the last century, we have experienced unprecedented losses of natural lands; yet it is the beauty of these natural landscapes that inspires many people to spend their lives here. When William Brewer, botanist with the California Geologic Survey, toured California in the early 1860s, the population of Los Angeles was about 4,000; by the mid-1990s it was almost four million. Who could have imagined the changes a little over a century would bring?

By the year 2005 the state's population is projected to increase to thirty-nine million people from today's thirty-one million. What can we do to ensure that our diverse floral riches and breathtaking landscapes remain for the benefit of future generations?

Citizen involvement has played a key role in the development of public policy to protect California's native plants and their habitats. Individual citizens and conservation organizations have had a strong influence in shaping California's conservation history. Scientists, teachers, landowners, legislators, agency personnel, conservation groups, and community members all have the ability to contribute to the conservation of our native plants and plant communities.

TRACKING PLANTS AT RISK

Much of our knowledge about California's rare plants began with an effort initiated in 1968 by G. Ledyard Stebbins, then professor of genetics at the University of California, Davis, and president of the California Native Plant Society. Dr. Stebbins organized a statewide compilation of the state's rarest plants. Since that time, professional and amateur botanists have worked together to document and map rare plant populations throughout California. In 1974 the California Native Plant Society published an *Inventory of Rare and Endangered Vascular Plants of California*, and, in 1994, its fifth and updated edition of the inventory was accompanied by an *Electronic Inventory* to provide computer access to current information. The fifth edition summarizes distribution, rarity, endangerment, and ecological information for 1,742 rare plants.

In 1979, at the urging of CNPS and The Nature Conservancy, the California Resources Agency established within the Department of Fish and Game a Natural Diversity Data Base (NDDB). This data base is a computerized inventory of the locations and status of the state's rarest species and natural communities. Today, the NDDB is one of California's most visible and successful tools for rare plant conservation. Its information is used by government agencies and the private sector in making land use and natural resource management decisions. The success of the NDDB rare plant and natural communities programs depends on data received from botanists and biologists throughout the state who submit information on plant population fluctuations, new locations, and the condition of rare plants and plant communities. Currently, the NDDB tracks about 1,750 plants of conservation concern and 135 rare plant communities.

HABITAT PROTECTION IS KEY

Conserving species in their natural setting, their own habitat, is essential to ensure their long-term survival. A plant in a botanic garden does not meet the conservation goal of preserving California's natural floristic diversity.

Habitat for rare plants on private lands can be protected through a variety of formal and informal means. The simplest way is when a landowner makes a personal commitment to care for plant habitat; many knowledgeable and concerned landowners in California are doing this. In some cases, private landowners may want to transfer management responsibility for a plant population and its habitat to someone else by either transferring an easement or the land itself, or by establishing a cooperative management agreement with a public agency, private conservation group, or an organization with experience in managing natural lands, such as a local land trust.

Throughout California, partnerships between state, federal, and local agencies, conservation organizations, land trusts, and private citizens successfully protect, maintain, and enhance plant populations and natural communities. California's voting public has approved government acquisition of important habitats through bond acts. One of these, Proposition 70, passed in 1988, has funded the acquisition of 36,000 acres of wildlands to date. In addition, Federal Land and Water Conservation funding has brought many threatened habitat areas into public ownership. Many natural areas have been set aside throughout California specifically to protect rare plant populations. Private and public owners and managers of these lands are protecting an important legacy for California's future.

A part of North Table Mountain in Butte County is a California Department of Fish and Game Wildlife Area.

MULTIPLE-SPECIES CONSERVATION AND LOCAL LAND USE PLANNING

Enacted in 1970, the California Environmental Quality Act (CEQA) requires state and local government agencies to identify and disclose environmental impacts of proposed projects and to consider ways to avoid or mitigate them where possible. Through the Environmental Impact

Report (EIR) process, the public can review proposed projects and attempt to influence decisions through public comment.

The most effective conservation for species and habitats comes from protection of large areas, such as whole watersheds or ecosystems. Such efforts help to ensure that viable natural systems are left intact to support the variety of wildlife and plants typical of the region. Well designed conservation areas often are more economical to manage in the long term because they are more self-sustaining than smaller preserves, which are easily disturbed. State, federal, and local agency efforts are increasingly focusing on regional con-servation planning.

In 1991, under the Natural Communities Conservation Planning Act, a pilot conservation planning program was initiated for the Southern California coastal sage scrub plant community. A similar but smaller conservation plan for Sonoma County's Santa Rosa Plain is being developed to protect the region's vernal pool ecosystem, home to several endangered plants.

The goal of regional planning efforts such as these is to protect areas large enough to promote the continued existence of multiple species and their habitats, while allowing for urban growth. Because many rare plants have narrow distributions, large-scale conservation efforts focused on natural communities must be carefully planned so that rare plant habitat is included within established conservation areas.

STEWARDSHIP AND RECOVERY PROGRAMS

Once habitat is legally protected, long-term stewardship of the land is essential. This entails monitoring rare plant populations over time to assess their health and detect potential threats. At the Lanphere-Christensen Dunes near Eureka, for example, The Nature Con-servancy (TNC) is re-storing a fragile dune ecosystem and moni-toring the return of native plants. Management activities there include removal of invasive exotic plants, peri-odic controlled burning, fencing to prevent damage from vehicles, stray pets, or livestock, and ecological restoration to repair previous damage. Volunteers contribute labor and expertise.

The traditional goal of plant conservation efforts is for endan-gered species to recover to a population size where they can flourish without human intervention. A more realistic goal may be to protect a declining species' habitat so it will con-tinue to thrive with minimal intervention. Recovery programs must be tailored to the needs of the species, using the most effective and efficient tools available. Efforts that promote collaboration and integrate land conservation actions with scientifically based stewardship and public outreach programs are often the most successful. Early prevention can be the key to success because efforts undertaken before a species reaches the brink of extinction are usually less costly and more effective.

The Nature Conservancy has maintained a weed removal and dune restoration program at the Lanphere-Christensen Dunes Preserve in Arcata for many years.

Protection of natural habitat and populations is essential; however, for many endangered species, it is also appropriate to store genetic material such as seeds or other reproductive parts at an off-site facility to provide additional insurance against extinction.

RESEARCH

For many species scientific knowledge needed for sound management is lacking. Academic researchers at universities, museums, botanical gardens, and private foundations work to answer questions that may be important to the conservation of rare plant populations. Research by professors and students of botany focus on plant population genetics, reproductive strategies of plants, long-term population trends, habitat characterization, and other topics that develop new ways to manage and conserve native plants. Examples of ongoing research include understanding Southern California's rare alluvial fan sage scrub plant community, identifying reproductive strategies of the endangered slender-horned spineflower, and genetic studies of Santa Cruz Island bush mallow. These studies and others guide management and recovery decisions for native plants and habitats known to be at risk.

SPECIES LISTINGS

Scientists, government agencies, citizens, and conservation organizations can recommend plants known to be at risk for official listing as rare, threatened, or endangered under the California Endangered Species Act or the federal Endangered Species Act. Today, 216 plants are officially listed as rare, threatened, or endangered by the State of California; seventy-six California species are officially listed as threatened or endangered by the federal government. Additional species continue to be proposed for listing. These designations provide essential public recognition of a species' plight and the need to conserve its habitat. Official listing can also require government agencies to protect species on public lands and provide increased consideration of species' needs when they occur on private lands. Western lily (*Lilium occidentale*), native to California's north coast and Oregon, is an example of a plant that has benefited from state and federal listing as an endangered species. Increased visibility has led to voluntary stewardship by private landowners, habitat protection and monitoring by local CNPS and TNC members, and habitat acquisition and restoration by TNC, the California Department of Fish and Game, and federal agencies.

Most California occurrences of the very rare western lily (Lilium occidentale) *are actively protected by private landowners.*

PUBLIC SUPPORT IS ESSENTIAL

Every resident and visitor to California can help conserve our dwindling native plant habitats. Involvement by concerned citizens, and groups such as the California Native Plant Society, is the essential ingredient for effective plant conservation. Public involvement affects the strength of our laws, the actions of public and private landowners to conserve our botanical heritage, and the ability of government agencies, charged with protecting nature, to follow through. Moreover, citizen involvement often swings the pendulum of decision-making toward a conservation solution. Improved government effort on behalf of plant conservation **is dependent on the** active role of private citizens.

Members of the California Native Plant Society lead plant hikes all over the state so people can learn about plants and enjoy the California landscape.

One of the most important steps individuals can take is to learn about native plants and habitats in their local area and to share their knowledge with others. Many who care deeply about the natural world have not yet learned about the riches in their own "backyard" or how they, as citizens, can influence conservation efforts. By visiting local natural areas and joining organized field trips, people can begin to become involved. They may decide to adopt a rare plant site and monitor it, attend meetings of local conservation groups, or become active in local land use planning issues.

Organizations such as the California Native Plant Society and The Nature Conservancy, government agencies such as the Department of Fish and Game and the Department of Parks and Recreation, and smaller jurisdictions such as counties, cities, and water and parks districts need assistance and expertise from local citizens. Their success depends on people who volunteer to remove exotic plants from native habitats, lead field trips, participate in restoration projects, develop local floras, monitor rare plants, and conduct research and education programs. Volunteers are the primary reason many conservation actions are successful.

Success in conserving native plants will depend on our collective creativity and conviction. Each of us can help in some small way to preserve California's rich botanical heritage.

Ken Berg, Sandra Morey, and Diane Ikeda

NORTH COAST AND KLAMATH RANGES

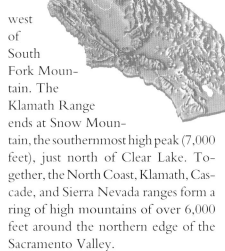

The northwestern region of California is perhaps best known for its majestic redwood forests and its rich diversity of conifers—seventeen species grow on the Salmon Mountains in the Russian Peak Area. Outcrops of serpentine rock are common throughout the region, and splendid examples can be seen along the Smith River, the Lassics, and Red Mountain in the North Coast Ranges.

Covering a vast area, the rugged mountains of the North Coast and Klamath ranges stretch eastward from the dramatic edge of the Pacific across to the volcanic mountains of the Cascade Range. From the Oregon border, the region reaches south across the deep scenic valleys of the Smith, Klamath, and Trinity rivers and ends in Mendocino County at the gentle Eel and Noyo rivers, and Lake County to

the east. In all, some 3,400 species of plants grow in the North Coast and Klamath region—half of those found in the entire state. And a surprisingly high number—285 species—occur nowhere else.

The Klamath Range, extending in a triangle south from Oregon, is a complex assemblage of mountains that include the Siskiyou, Scott Bar, Marble, Salmon, Trinity Alps, and Trinity Mountains south to the Yolla Bolly Mountains in northern Mendocino County. Abutting the North Coast Ranges to the south, the Klamaths are older, more rugged, and taller, reaching over 9,000 feet at Mount Eddy. The North Coast Ranges widen south of the Klamath Range, from which they are separated by a geological fault

west of South Fork Mountain. The Klamath Range ends at Snow Mountain, the southernmost high peak (7,000 feet), just north of Clear Lake. Together, the North Coast, Klamath, Cascade, and Sierra Nevada ranges form a ring of high mountains of over 6,000 feet around the northern edge of the Sacramento Valley.

Steep slopes and headlands line the coast. Near Cape Mendocino, King Peak rises directly out of the sea to 4,000 feet. In a few places along the coast ancient marine terraces remain, some with as many as five identifiable steps, as at the Jughandle State Reserve Ecological Staircase near Fort Bragg. The lowest and youngest terraces here are covered with sand as at Humboldt Bay and Point George. Though less than ten percent of the historic marshes remain today in Humboldt Bay, the bay continues to shelter almost all the salt marshes of the northwest region.

Along the North Coast rare plant habitat often occurs near population centers and is eliminated by urban expansion. In more rural areas logging, grazing, road building, and off-road vehicle use threaten numerous plant species. The once extensive forest has been changed by decades of logging, and today there is considerable controversy as to how to best manage the remaining old-growth forests.

John Sawyer

Along the North Coast, coast redwoods grow where summers are cool and foggy and winters are warm, in strong contrast to hot summers and deep snow inland. Moist redwood forests are rich in California rose-bay (Rhododendron macrophyllum), RIGHT, *and western azalea* (R. occidentale) *found growing near streams. Uncommon floral treasures found in this corner of the state include California lady's-slipper* (Cypripedium californicum), LEFT, *and yellow skunk cabbage* (Lysichiton americanum), ABOVE.

The North Fork of the Smith River, OPPO-SITE, *is a national recreation area, managed by Six Rivers National Forest. Legislation establishing the recreation area prohibits new mining claims, though existing claims will be honored. Twenty-one rare and endangered plants, such as McDonald's rock cress* (Arabis macdonaldiana), *Waldo buckwheat* (Erio-gonum pendulum), *and Howell's jewel-flower* (Streptanthus howellii), *grow on the extensive serpentine rock outcrops that extend northward into Oregon. • Uncommon species such as yellow-flowered iris* (Iris chrysophylla), LEFT, *redwood lily* (Lilium rubescens), BELOW, *and the Klamath fawn lily* (Erythronium klamathense), BOTTOM, *grow in light gaps in the forest canopy.*

NORTH FORK OF THE SMITH RIVER

The most pristine river system in California, the Smith River is the only major un-dammed river remaining in California. The watershed of the North Fork of the Smith River is known for its rugged topography, the highest recorded rainfalls in the state (up to 120 inches annually), and high concentrations of rare plants.

Dwarfed, fire-adapted knobcone pine (*Pinus attenuata)* and lodgepole pine (*P. contorta*) grow on the upper slopes of the North Fork, while Jeffrey pine (*P. jeffreyi*) woodlands dominate the rocky ridgelines below 3,400 feet. Forests at middle to lower elevations are composed largely of Douglas-fir (*Pseudotsuga menziesii*) and incense cedar (*Calocedrus decurrens*). The rare Port Orford cedar (*Cupressus lawsoniana*) and its common associate, western azalea (*Rhododendron occidentale*), ribbon the tributaries and the main stem of the Smith's North Fork at about 600 feet elevation.

Because of the biological significance of the area and its concentration of rare plants, the U.S. Forest Service

(USFS) has proposed 1,305 acres of seep habitat as a research natural area to encourage research and education. The USFS has also proposed a majority of the North Fork watershed as a North Fork Smith Botanical Area to conserve its unique botanical values and to provide recreational opportunities compatible with those values.

While these designations afford some protection for the North Fork of the Smith River, the plants and habitat of the watershed remain vulnerable to losses from mining and related road construction, recreational activities, Port Orford cedar root disease, and even horticultural collecting.

Lisa D. Hoover

LANPHERE-CHRISTENSEN DUNES

Near the North Coast city of Eureka, the 450-acre Lanphere-Christensen Dunes Preserve protects the most pristine coastal dunes remaining in the Pacific Northwest. These dunes are a remnant of a large coastal dune system, one of several in California. Typically, coastal dunes form as sand erodes from inland mountains and travels to the sea along coastal rivers. Currents and waves transport the sand onto beaches where it is shaped into dunes. For the Lanphere-Christensen Dunes the source of sand is the free-flowing Mad and Eel rivers.

In sheltered depressions, dune hollows support marshes populated by sedges, rushes, and, in the more mature hollows, groves of native willow and beach pine (*Pinus contorta*). Farthest from the beach on older, more stable dunes, a beach pine and sitka spruce (*Picea sitchensis*) forest grows. Creeping bearberry (*Arctostaphylos uva-ursi*) and reindeer lichen carpet the

To survive wind, sea spray, shifting sand, and seasonal drought, dune mat plants have special adaptations—prostrate growth forms, large fleshy taproots, and fleshy, succulent leaves as shown by beach morning-glory (Convolvulus soldanella) *and yellow sand verbena* (Abronia latifolia), TOP. • *To deflect excessive amounts of sunlight, beach pea* (Lathyrus littoralis), ABOVE, *has evolved leaves with fine white hairs.* • *The bright yellow endangered Humboldt Bay wallflower* (Erysimum menziesii *ssp.* eurekense), OPPOSITE PAGE, TOP LEFT, *is one of the many beautiful wildflowers covering the dunes.* • *The native dune grass* (Poa douglasii) *shown here,* OPPOSITE PAGE, MIDDLE, *has been nearly eliminated by aggressive European beachgrass* (Ammophila arenaria). • *A diverse and unique plant community, known as dune mat,* OPPOSITE PAGE, BOTTOM RIGHT, *is found on the semi-stable foredunes nearest the sea. Composed of more than fifty species, dune mat offers spectacular spring and summer color.*

WESTERN LILY

One of the most spectacular plants native to California, the rare western lily (*Lilium occidentale*), grows only at a few locations near the coast between Humboldt Bay and Coos County, Oregon. As recently as 1980 the western lily was believed to be perilously close to extinction from more than a century of farming and grazing, development, and horticultural bulb collection.

The outlook for the western lily, which is listed as an endangered plant in both Oregon and California, has brightened in the past decade. Botanists have discovered or rediscovered several populations of the lily, volunteers are monitoring known populations, and several sites have been acquired to protect lily habitat. The California Department of Fish and Game has acquired the two largest known populations, at the Table Bluff Ecological Reserve near Eureka, while The Nature Conservancy has acquired additional critical habitat near Coos Bay, Oregon. Both reserves are now managed to preserve the lily. Annual monitoring and habitat restoration have been funded through contributions to the California Endangered Species Tax Check-Off Fund. Research is underway to examine long-term effects of controlled burning, grazing, and other habitat manipulation.

Dave Imper

As recently as fifteen years ago, western lily (Lilium occidentale) *was known from only one population in California. Several other populations have been discovered since, and some are now protected and thriving.*

floor of the forest, where an unusual coastal population of the beautiful pink calypso orchid (*Calypso bulbosa*) adds a welcome spot of color.

The Lanphere-Christensen Dunes Preserve was established in 1974 by The Nature Conservancy, which later expanded the preserve's borders by incorporating less pristine dunes that had been damaged by off-road vehicles and encroaching introduced species. An ambitious, long-term program initiated to restore these dunes is nearing completion. The first task, a somewhat daunting one, was to remove European beachgrass (*Ammophila arenaria*), an aggressive introduced species commonly used for stabilizing dunes. The restoration is showing success, and native plants such as the endangered beach layia (*Layia carnosa*) are returning to many areas of the dunes.

Andrea Pickart

HUMBOLDT BAY SALT MARSHES

Coastal development and agricultural reclamation have resulted in a dramatic decline in Northern California salt marsh habitat. Of the more than 7,000 acres of salt marsh that once bordered Humboldt Bay, only about 900 acres remain today. This remnant fringe, occurring along the edges of the bay and its islands, is the principal site of salt marsh between San Francisco Bay and Coos Bay, Oregon.

At lowest elevations, just above the mud flats of Humboldt Bay, the most common plant is pickleweed (*Salicornia virginica*). Cordgrass (*Spartina foliosa*) is sparse or largely absent, while non-native Chilean cordgrass (*S. densiflora*) has spread in mid-range elevations so extensively that it has become a dominant species in salt marsh vegetation. At high elevations above the high-tide line, the dominant species are pickleweed, jaumea (*Jaumea carnosa*), and saltgrass (*Distichlis spicata*).

Two members of the figwort fam-

ily (Scrophulariaceae) that occur in the upper marsh are considered rare: Humboldt Bay owl's-clover (*Castilleja ambigua* ssp. *humboldtiensis*) and the hemiparasitic Point Reyes bird's-beak (*Cordylanthus maritimus* ssp. *palustris*). Humboldt Bay owl's-clover is known only

from Humboldt Bay and Point Reyes; Point Reyes bird's-beak ranges from Coos Bay, Oregon, to San Francisco Bay. Both are endangered by loss of habitat and coastal development, but are protected within the Humboldt Bay National Wildlife Refuge of the U.S. Fish and Wildlife Service and The Nature Conservancy's Lanphere-Christensen Dunes Preserve. Sound habitat management will be critically important for the long-term survival of these two species now confined to the margins of their original habitat.

Anni L. Eicher

The rare Point Reyes bird's-beak (Cordylanthus maritimus *ssp.* palustris), UPPER LEFT, is found only in high-elevation marshes since Chilean cordgrass (Spartina densiflora) has become a dominant species in the bay. • Brass-buttons (Cotula coronopifolia), ABOVE, native to South Africa, grows along with cattails (Typha *sp.*) in a ponded high marsh area. • Chilean cordgrass (Spartina densiflora), LEFT, was introduced into Humboldt Bay in the mid-1800s by lumber ships from Chile. • Vegetation grows lush in North Coast salt marshes where rainfall reduces salinity levels. Common fire-weed (Epilobium angustifolium *ssp.* circumvagum), FAR LEFT, is in the foreground.

COAST REDWOOD

California's coast redwood (*Sequoia sempervirens*) is the last relict of a group of *Sequoia* species that were part of an ancient temperate forest encircling the northern hemisphere roughly between the latitudes of Alaska and Oregon. Fifty million years ago, these fossil redwoods mixed with a rich array of other conifers—ginkgo, dawn redwood, giant sequoia, bald cypress, fir, hemlock—and winter-deciduous hardwoods—alder, basswood, beech, chestnut, elm, hickory, maple, sassafras, sweetgum, and oak. The climate then was warm and humid in summer, mild and damp in winter.

As the climate cooled and dried, suitable redwood habitat shrank. The mixture of trees fragmented and became less rich. All but one redwood species became extinct. By the onset of the Ice Age, two million years ago, the one remaining species had become

restricted to the Pacific coast of North America. Cycles of glacial advance and retreat, accompanied by increasing summer drought, eventually limited coast redwood to its present sliver of earth, 450 miles long, never more than thirty-five miles from the ocean, and below 3,000 feet in elevation.

Although coast redwood suffers under the stresses of frost, heat, and drought, it is adapted to tolerate fire and flood. It withstands the heat from ground fires because its bark is thick and fibrous, providing insulation to living tissue within. After a fire, if a tree is significantly burned, dormant buds in the root-stem (burl) region are stimulated; they sprout and push through the bark and emerge above the soil. Sprouts grow rapidly because of their attachment to the enormous root system of the parent, already in place. In time, some of the sprouts become mature trees, encircling the empty space where the parent once grew.

Two centuries ago, California had 1.8 million acres of redwood forest; today, there is still nearly the same acreage, but it is not old-growth. About 50,000 acres of old-growth forest remain, but only 39,000 acres are protected in state and federal parks. The unprotected remainder may be logged by the millennium, leaving a heritage of two percent of what the state once had, and of this little or no old-growth forest.

Michael Barbour and Valerie Whitworth

Douglas iris (Iris douglasiana), OPPOSITE, BOTTOM LEFT, *giant trillium* (Trillium chloropetalum), OPPOSITE, TOP, *foetid adder's tongue* (Scoliopus bigelovii), OPPOSITE, MIDDLE, *and western wood anemone,* (A. oregana), LEFT, *often grow on the moist redwood forest floor where a giant redwood has fallen.* • Redwood trees, ABOVE, *commonly attain heights over 300 feet, trunk diameters greater than ten feet, ages over 1,000 years, and weights approaching 500 tons (two to three times heavier than the largest animal that ever lived, the blue whale). These leviathans live only in old-growth forests, which have a unique architecture and complement of species. Certain fungi, shade-requiring plants, wood-eating insects, and burrowing animals require old-growth forests for normal life and reproduction, and they are not found in recovering forests. Redwood lumber has been harvested extensively since the early 1800s. Clear-cutting is the usual harvest method. Thirty to forty acres are logged at a time and then the area is allowed to recover. Redwood stumps are left in place to resprout. Both redwood and Douglas-fir grow quickly and produce trees large enough to harvest again within seventy years, but these second-growth forests are not the equivalent of old growth.*

THE YOLLA BOLLY MOUNTAINS

The remote Yolla Bolly Mountains, with several peaks over 7,000 feet, lie immediately west of the northern Sacramento Valley between Willows and Red Bluff. Because the Yolla Bolly Mountains are the tallest in the North Coast Ranges and have had relatively direct connections with the boreal regions of the Klamath Ranges, many montane and subalpine plants (well over 200 species) find their southernmost coastal range limit here. These include foxtail pine, mountain hemlock, quaking aspen, Jeffrey pine, and scattered stands of western juniper and western white pine.

Such characteristic subalpine species as purple round-leaved buckwheat (*Eriogonum ovalifolium* var. *purpureum*), luetkea (*Luetkea pectinata*), blue sky pilot (*Polemonium pulcherrimum*), and violet rock penstemon *(Penstemon rupicola)* reach their southern limits in the highest, northern portion of these mountains. Shasta red fir, simple grapefern (*Botrychium simplex*), mountain maple (*Acer glabrum* var. *torreyi*), and western aster *(Aster occidentalis)* occur in the vicinity of Snow Mountain, the southernmost of the 7,000-foot peaks. Here botanists Larry Heckard and James Hickman found 126 species that extend no farther south in the Coast Ranges.

The Anthony Peak lupine (*Lupinus antoninus*), Stebbin's lewisia (*Lewisia stebbinsii*), Snow Mountain buckwheat (*Eriogonum nervulosum*), and Snow Mountain willowherb (*Epilobium nivium*) have limited populations and are notably rare. At least two rare species, Yolla Bolly Mountain bird's-foot trefoil (*Lotus yollabolliensis*) and clustered green-gentian (*Swertia fastigiata*), are locally restricted to the South Fork Mountain schist running from North Yolla Bolly northwest to Pickett Peak.

The remoteness of this area and the fact that most land in the vicinity of North and South Yolla Bolly is in the Shasta-Trinity, Mendocino, or Six Rivers National Forest bode well for the conservation of the region's plant resources.

Todd Keeler-Wolf

In the language of the Wintu Indians, Yolla Bolly means "high snow-covered peak," and indeed, the highest peaks in this range, ABOVE RIGHT, *retain snowbanks well into summer most years. Because of generally lower elevations and a more southerly location, there are fewer lakes, wet meadows, and permanent streams here than in the high Klamaths and summers are hot and dry. • The distinctive Snow Mountain beardtongue* (Penstemon purpusii), ABOVE LEFT, *Hoover's lomatium* (Lomatium ciliolatum *var.* hooveri)*, and scabrid raillardella* (Raillardiopsis scabrida) *grow only in this mountain range but are locally widespread. •* Lewisia stebbinsii, BELOW, *is found on the dry, rocky slopes of these mountains.*

RED MOUNTAIN

The pronounced red color of Red Mountain in Mendocino County, a 4,000-foot-high rounded peak, signals the presence of significant nickel and chromium deposits that are of considerable interest to both private mining interests and to botanists drawn to the unique soils. Located three miles northeast of Leggett, California, Red Mountain is composed of highly weathered soils derived from ultramafic rock and is home to several rare plants.

The vegetation of Red Mountain is largely open woodland with sugar pine (*Pinus lambertiana*) and other conifers and a sparse understory of manzanita, silktassel, and other low shrubs. The area is remarkably free of weedy introductions.

Most of the crest of the mountain, which is home to many of the rare and endangered plant populations, is managed by the Bureau of Land Management (BLM). Following the listing of McDonald's rock cress (*Arabis macdonaldiana*) as an endangered species by federal and state governments in the late 1970s, much of the surrounding private land passed from the hands of a mining company to a timber-growing concern.

In 1984 BLM initiated a long-term monitoring program to study the population dynamics, geographic distribution, and habitat requirements of the four plant species endemic to Red Mountain. BLM will track patterns of change within populations of each species and manage for the plants' long-term survival on the site.

Mike Baad

Evening fog surrounds mixed conifers on Red Mountain, ABOVE. • Red Mountain stonecrop (Sedum eastwoodiae), BELOW, is one of four rare narrowly endemic plant species growing on windswept ridges and talus slopes where there is little competition from other species.

PYGMY FOREST OF MENDOCINO

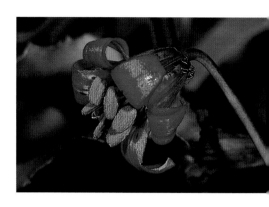

A pygmy forest grows on the oldest of the uplifted ancient marine terraces on the Mendocino County coast. These terraces provide one of the exceptional soil and vegetational features of the California coastline. A staircase sequence of terraces, youngest at the water's edge and progressively older away from the ocean, has formed through geologic uplifting over many thousands of years.

Adjacent areas that support towering redwood forests share greywacke sandstone bedrock material and a foggy, cool climate; the difference lies in the soil: pygmy forest soil is very old. It has had thousands of years of rainfall to leach out nutrients and minerals and coniferous cover to acidify it. A concrete hardpan lying approximately eighteen inches below the soil surface has formed as a result of iron compounds in the highly acidic soils. Nearly a million years of leaching of soils on the most ancient, thus most inland, terraces has created a white, nutrient-poor, iron-hardpan soil, sometimes called the White Plains of Mendocino, which sustains only stunted vegetation—a pygmy forest—where trees, hundreds of years old, often grow only two or three feet high. Once there were 4,000 acres of pygmy forest on remnants of the most ancient terraces. More than half have been severely degraded by garbage dumps, airports, houses, and off-road vehicle traffic. Only isolated segments of this famous forest are protected in Van Damme State Park and Jughandle State Ecological Staircase. The Nature Conservancy and the University of California Natural Reserve System each hold a few small remnants. Mendocino County is required by a court order to classify pygmy forest lands as an Environmentally Sensitive Habitat Area (ESHA) in their county general plan, which should result in further protection of this rare and unusual habitat.

Teresa Sholars

Professor Hans Jenny at the University of California, Berkeley, identified the Mendocino pygmy forest area, RIGHT, *as one of the only places on earth where ocean terraces have remained flat as they have been uplifted. On the upper terrace two rare species grow, the pygmy cypress* (Cupressus goveniana *ssp.* pigmaea), *the smallest cypress tree in the world and a pygmy shore pine called Bolander pine* (Pinus contorta *ssp.* bolanderi). • *The beautiful deep red coast lily* (Lilium maritimum), ABOVE, *creates bright flecks of color in gaps in the pygmy forest.* • *Western azalea* (Rhododendron occidentale), BELOW, *grows in moist soils near streams.*

BEAR VALLEY

Spring brings a spectacular wild-flower display to the southern end of Bear Valley in Colusa County. Located west of Williams and east of Clear Lake, Bear Valley lies at an elevation of approximately 1,300 feet and is bounded on the west by Walker Ridge. By April and early May thick sheets of wild-flowers spread across the valley floor, reminiscent of the mid-nineteenth century when vast areas of the Great Valley were so densely carpeted with wild-flowers that "an ant could walk from flower to flower for miles, never touching ground."

People from all over Central and Northern California make a yearly pilgrimage to see the show, a landscape painted with wildflowers. In mid-March the floor of the southern valley is covered with hundreds of thousands of pink adobe-lilies (*Fritillaria pluriflora*). By April some of the most stunning fields are living tapestries of pale cream cups (*Platystemon californicus*), bright orange poppies (*Eschscholzia californica*), purple owl's-clover (*Castilleja exserta* ssp. *exserta*), and two species of blazing yellow tidy-tips (*Layia* spp). The rare large-flowered star-tulip (*Calochortus uniflorus*) is locally abundant.

The valley's floral richness can be attributed to its soils but also to the fact that currently both plowing and livestock grazing have been limited voluntarily to a level that has helped to pre-

In March hundreds of thousands of adobe-lilies (Fritillaria pluriflora) *emerge,* TOP, *probably the most extensive display of any fritillary in North America. Grazing, off-road vehicles, and horticultural collecting all affect the population of adobe-lilies. • By April the floor of Bear Valley,* RIGHT, *is covered with tidy-tips* (Layia spp.) *and California poppy* (Eschscholzia californica), ABOVE. *This is one of the grandest display of lowland field wildflowers remaining north of the San Francisco Bay Area.*

serve spring displays. Outside pasture fences, exotic grasses and weeds tend to dominate.

Development pressures currently threaten this last vestige of spring floral profusion, which is in private owner- ship. One recent proposal would turn the valley floor into a large retirement complex, complete with vineyards, golf courses, tarmac, and ranchettes; an- other would develop irrigated pastures for intensive year-round grazing. Cali- fornians concerned about the future of this area are encouraging ranchers to preserve this rich remnant of Great Valley flower fields for future genera- tions.

Stephen W. Edwards

BOGGS LAKE

oggs Lake, southwest of Clear Lake in Lake County, is one of the largest vernal pools in Northern California. Nestled in the coastal mountains at 2,000 feet elevation in a basin of volcanic rock, Boggs Lake dries almost completely each summer, creating a giant ninety-acre vernal marsh that combines char-

acteristics of freshwater marshes and vernal pools. Boggs Lake's unusual setting has made it a refuge for several rare plants, including a few endemic species.

Distinctive throughout the year, the shallow bowl-shaped lake is especially beautiful in spring and early summer. The lowest areas are usually under water year-round and are ringed by tall tules and cattails typical of shallow ponds and freshwater marshes. Early in spring, when evaporating lake waters become warm and shallow, beautiful floating and submerged aquatic plants appear: common pondweeds (*Potamogeton* spp.) with reddish floating and narrower submerged leaves; arrowhead (*Sagittaria cuneata*) with leaves shaped like arrowheads and emergent white flowers; watershield (*Brasenia schreberi*); water-milfoil (*Myriophyllum hippuroides*), closely related to a common aquarium plant; and short-spurred bladderwort (*Utricularia gibba*) with emergent yellow flowers resembling snapdragons.

On the emerging shore, a succession of tiny annual wildflowers forms colorful rings similar to those found in smaller vernal pools. Here are found the rare vernal pool grass (*Orcuttia tenuis*), the tiny mud-dwelling figwort relative Boggs Lake hedge-hyssop (*Gratiola heterosepala*), exquisite lobelia-like downingias (*Downingia bicornuta*, *D. pulchella*, and *D. cuspidata*) with tiny blue-lipped flowers; and two rare navarretias (*Navarretia leucocephala* ssp. *pauciflora* and *N. leucocephala* ssp. *plieantha*), the latter endemic to this area.

Loch Lomond button-celery (Eryngium constancei), ABOVE, *grows in only one vernal pool, at nearby Loch Lomond. The California Department of Fish and Game purchased this vernal pool in the mid-1980s and created the Loch Lomond Ecological Reserve.*

Boggs Lake and its surrounding ponderosa pine forest are now protected by The Nature Conservancy, which manages the preserve and monitors rare plants.

Glenn Keator

In spring, downingias (Downingia spp.) *and the rare many-flowered navarretia* (Navarretia leucocephala *ssp.* plieantha), OPPOSITE, TOP, *emerge in a colorful band along the shores of the drying lake. • By late spring,* LEFT, *thick vegetation surrounds deeper parts of the vernal pool. • Boggs Lake hedge-hyssop* (Gratiola heterosepala), ABOVE LEFT, *grows here and in other scattered vernal pool and lake margin habitats in California and Oregon. • Flowers of the unusual mud plant, watershield* (Brasenia schreberi), ABOVE CENTER, *emerge twice, once to expose the stigma for pollination, and the next day to shed pollen. • Bladderworts, such as this rare short-spurred bladderwort* (Utricularia gibba), ABOVE RIGHT, *have underwater insect larvae-trapping bladders among their highly divided leaves.*

RUSSIAN PEAK

The largest concentration of cone-bearing trees in North America—seventeen in all—flourishes above 4,500 feet on the eastern flank of Russian Peak in Siskiyou County, where Sugar Creek drains the granitic Salmon Mountains. Such remarkable diversity probably results from geologic and climatic conditions over the past 10,000 years that allowed many conifer species from the Sierra Nevada, Klamath, Cascade, and Coast ranges to migrate into this area and flourish.

A walk to Sugar Lake from State Highway 3 rewards the hiker with breathtaking panoramic views of the glacial Duck Lake to the north, Russian Peak and the Trinity Alps to the south, and Mount Shasta to the east. Along the way, the hiker first encounters Shasta fir (*Abies magnifica* var. *shastensis*), incense cedar, Brewer spruce (*Picea breweriana*) and lodgepole pine. As elevation increases, there are Jeffrey pine, sugar pine, western white pine, ponderosa pine, Douglas-fir, white fir, and mountain hemlock. About halfway to the lake, away from the trail, are the creekside subalpine fir (*Abies lasiocarpus*), Engelmann spruce (*Picea engelmannii*), and western yew (*Taxus brevifolia*). Up the slopes to the ridge between Sugar Lake and Duck Lake creeks, ground juniper (*Juniperus communis*), whitebark pine (*Pinus albicaulis*), and foxtail pine (*P. balfouriana*) all survive the harsh conditions of snow and wind.

The U.S. Forest Service has designated Russian Peak and its drainage as part of Sugar Creek Research Natural Area for scientific study in the Klamath National Forest and for protection and management of the forest.

John Sawyer

PORT ORFORD CEDAR

The once majestic, old-growth stands of Port Orford cedar (*Cupressus lawsoniana*) in Northern California and southern Oregon are fast disappearing because of an exotic and fatal root rot, *Phytophthora*. Port Orford cedar rarely occurs in pure stands and is more often associated with Douglas-fir, western hemlock, coast redwood, or a variety of other conifers. This cedar grows on a remarkable range of parent materials and soils and over a broad elevational range, which is puzzling considering its highly restricted distribution. A requirement for moist soil through the summer seems to explain the eastern and southern range limits. The northern limit may be related to length of the growing season, fire frequency, or perhaps soil chemistry.

Faced with increasing public alarm over the severity of the disease, and a potential lawsuit in the 1980s, the U.S. Forest Service implemented a coordinated response involving both public agencies and the private sector. An action plan was developed in 1988 to monitor the disease, promote research, educate and involve the public, and coordinate management among affected landowners. Methods now used to slow the spread of the disease include removal of the host cedar in infested or threatened areas, sanitation of vehicle wheels, drainage control, and limited road closures. The success of these programs is uncertain.

Dave Imper

Now one of the rarest conifers in North America, Port Orford cedar, ABOVE, *is generally restricted to a forty-mile-wide coastal strip between Reedsport, Oregon, and Eureka, California, with a disjunct population in the upper Sacramento and Trinity river drainages. The* Phytophthora *fungus spreads into new areas primarily by spores on mud carried on vehicle wheels. Once in a watershed, the fungus can move rapidly through surface waters. Lowland and wetland stands are at greatest risk. The potential for loss of genetic diversity in Port Orford cedar could affect its long-term ability to adapt to altered climates. Even if an effective means of disease prevention is found soon, the unique and humbling experience of standing among old-growth Port Orford cedar will be missed by future generations.*

Seventeen species of conifers occur along the trail to Sugar Lake near Russian Peak, ABOVE LEFT. • *The handsome Wiggins lily (Lilium pardalinum ssp. wigginsii),* LEFT, *grows in wet places among the conifers.*

MOUNT EDDY

Mount Eddy, the eastern outpost of the Klamath Range where it meets the Trinity Mountains at the edge of the Cascade Range, boasts the largest outcrop of serpentine in North America. Overlooking the volcanic fields of the Cascades, Mount Eddy is home to wide-ranging Klamath Range endemics such as Copeland's speedwell (*Veronica copelandii*) as well as to local Trinity Mountain serpentine endemics, including the state-listed Trinity buckwheat (*Eriogonum alpinum*). The rare Klamath

Not all of the rare plants on Mount Eddy are alpine species. This showy raillardella (Raillardella pringlei), TOP, *inhabits the numerous darlingtonia seeps and streamsides on the mountain. • Several curious disjunct alpine plants also grow near the summit. Perhaps most intriguing is Mason's sky pilot (Polemonium chartaceum),* ABOVE, *also found on nearby Scott Mountain, but otherwise known only from the Sweetwater Mountains east of the Sierra Nevada. • Mount Eddy cryptantha (Cryptantha sobolifera),* RIGHT, *is found only on volcanic soils above 4,000 feet. • The handsome trunk of a foxtail pine (Pinus balfouriana),* FAR RIGHT, *frames this view of Little Crater Lake on Mount Eddy.*

manzanita (*Arctostaphylos klamathensis*) and white Scott Mountain phacelia (*Phacelia dalesiana*) grow in the understory of the upper montane mixed conifer forest on the west slope of Mount Eddy.

Mount Eddy is within the boundaries of the Mount Shasta Ranger District of the Shasta-Trinity National Forest. Logging and grazing, activities that threaten rare plant species, are part of current land use management of the area and have been for many years. The summit trail is heavily used by hikers because of its rich wildflower displays and the spectacular view of Mount Shasta from the top.

The Forest Service proposes establishment of a foxtail pine (*Pinus balfouriana*) research natural area at the summit of Mount Eddy, where rare plants such as Trinity buckwheat also grow. Through a land exchange, the Forest Service is attempting to acquire an important large and private inholding north of the summit. If successful, this transaction will bring into public ownership one of California's premier rare plant habitats.

Julie Kierstead Nelson

CASCADE RANGE AND MODOC

The Cascade Range and Modoc Plateau regions of northeastern California offer striking contrasts in natural features. From the densely forested western slopes of the Cascade Range to the dry, sagebrush-covered slopes spilling into the Great Basin to the east, this part of California is characterized by geological and botanical extremes. For all its diversity, the area remains one of the least known, botanically, of any in the state.

Mountains reaching to more than 14,000 feet dot the landscape in the western portion of the region. Near Susanville, the volcanic Cascade Range meets the northern end of the granitic Sierra Nevada. Precipitation comes predominantly as snow in the Cascades, and can reach more than eighty-five inches at upper elevations. The basalt rock that underlies the entire region has weathered to a moist acidic soil in the west, supporting alpine communities on Mount Shasta and Mount Lassen, and mixed conifer and ponderosa pine forests on the lower slopes. To the east, where the rainshadow of the western mountains creates drier conditions, mid-elevations support western juniper (*Juniperus occidentalis*), sagebrush steppe, and bunchgrass communities.

This northeastern portion of California is dominated by two major geologic features: the Modoc Plateau, an area of geologically young basalt flows and volcanic shields, which includes the western extent of the Great Basin; and the Warner Mountains, a north-south trending range near the Nevada border extending into Oregon. On the Modoc Plateau, soils are alkaline and precipitation can total as little as four

California Indian pink (Silene californica), LEFT, *a bright member of the pink family, occasionally decorates rocky or gravelly slopes in spring.* • *Two volcanic peaks, Mount Shasta*, RIGHT, *and Mount Lassen, dominate the southern end of the Cascade Range. Because of their young geology and low diversity of rock types and soils, the Cascade Range and Modoc Plateau regions have few rare plants and endemic species. Several plants more common in other states but rare in California occur on the Modoc Plateau, including silverleaf milk-vetch* (Astragalus argophyllus), *lavender blue Great Basin downingia* (Downingia laeta), *and purple-flowered dwarf lousewort* (Pedicularis centranthera). • *In early spring, hound's tongue* (Cynoglossum grande), BELOW, *named for its long tongue-shaped leaves, brightens the floor of moist, shady woodlands.*

PLATEAU

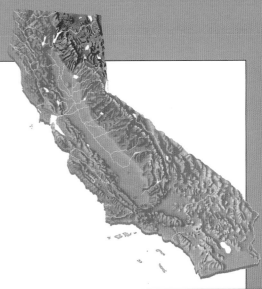

inches annually. Here, dry desert sage-brush communities dominate the landscape, though, when periodically inundated, shallow alkaline lakes also become prominent features. The Warner Mountain Range with peaks of nearly 10,000 feet supports a mix of Sierran and Great Basin floras, although two common Sierran conifers, red fir (*Abies magnifica*) and sugar pine (*Pinus lambertiana*), are absent. A high-elevation pine, the Washoe pine (*Pinus washoensis*), is common in the southern portion of the Warners and is also found on the eastern slopes at the northern end of the Sierra Nevada.

Gary Schoolcraft

MOUNT SHASTA

The magnificent snowy cone of Mount Shasta can be seen from almost half of California's Great Valley; at 14,162 feet, its peak is the sixth highest point in the state. Lying in the rain shadow of the Klamath Range, it was described in 1862 by William Brewer in his journal, *Up and Down California*, as "sublime in its desolation, Mount Shasta is among the driest of all the Cascade peaks." C. Hart Merriam noted in an 1899 biological survey of Mount Shasta that many characteristic montane plants of the Cascades and Sierra Nevada were unable to fill their usual elevational niches on the drier slopes of Mount

Two other local endemic plants, Cooke's phacelia (*Phacelia cookei*) and pallid bird's-beak (*Cordylanthus tenuis* ssp. *pallescens*), hug the dry, sandy skirts of the mountain. The bird's-beak is found on rural roadsides in the towns of Mount Shasta and Weed at the base of the mountain. Both of these annual species rely on timely soil disturbance for successful germination and completion of their life cycle.

Mount Shasta is entirely within the Shasta-Trinity National Forest. The Forest's Mount Shasta Ranger District recently developed a restoration plan for Panther and Squaw Creek meadows based on studies of the vegetation and patterns of human use. The Mount Shasta Wilderness Plan, which the district has nearly completed, also includes protection for the scarce riparian habitat in Squaw Creek Meadows.

Development of a ski area near Panther Meadows has been proposed many times but remains controversial.

Julie Kierstead Nelson

Shasta; instead, they were restricted to moist areas near springs and streams.

Dominant vegetation on Mount Shasta includes dense stands of greenleaf manzanita (*Arctostaphylos patula*) and bitterbrush (*Purshia* spp.). High elevations are dominated by pure stands of the stately Shasta red fir (*Abies magnifica* var. *shastensis*). The two prominent meadow systems, Panther and Squaw Creek meadows, are both home to a rare perennial herb, Wilkin's harebell (*Campanula wilkinsiana*), which has suffered as a result of the proliferation of trails through the meadows and diversion of stream water for campers.

Through a stand of western blue-flag (Iris missouriensis) on the shore of Grass Lake, the snowy peaks of Mount Shasta and Shastina rise like snow cones from the meadows below, LEFT. • *The uncommon clustered lady's-slipper (Cypripedium fasciculatum), shown here in bud,* FAR LEFT, *is threatened by logging and horticultural collecting, and many populations appear to be reproducing poorly.* • *Occasionally a visitor will find the handsome white flower of Washington lily (Lilium washingtonianum ssp. washingtonianum),* ABOVE LEFT, *in clearings of a coniferous forest.* • *Sulfur flower (Eriogonum umbellatum),* ABOVE, *grows on the lava beds of Mount Shasta, and in many other dry, open places.*

SHASTA SNOW-WREATH

The discovery of Shasta snow-wreath (*Neviusia cliftonii*), BELOW RIGHT, in 1992 ranks among California's most exciting and unexpected botanical finds of this century. Though growing in large stands next to a state highway and a U.S. Forest Service campground, the Shasta snow-wreath had been overlooked until two botanists, Glenn Clifton and Dean Taylor, were able to cross a creek in a drought year. There they found an unfamiliar plant, later identified as a previously undiscovered relict species and primitive member of the rose family. It is now known from eight populations in the Shasta Lake area; the only other species in the genus, Alabama snow-wreath (*Neviusia alabamensis*), occurs 2,000 miles away in the southeastern United States.

Barbara Ertter

LASSEN VOLCANIC NATIONAL PARK

Lassen Volcanic National Park provides 105,000 acres of extraordinarily rich habitat for plants owing to the region's broad elevational range and its location at the junction of two major mountain systems, the Cascade Range and the Sierra Nevada. Lassen Park has been called a miniature Yellowstone because, though lacking geysers, its unique geology features hissing steam vents, boiling pools and gurgling mudpots. These areas, uncommon in California, offer a good place to see plants adapted to warm water and unusual minerals. Approximately 800 plant species, subspecies, and varieties grow within the vicinity of Lassen Volcanic National Park, including more than twenty species reaching their northern limit of distribution in the park and a similar number of Cascadian species reaching their southern range limit within the park.

Lassen Peak, the southernmost major peak of the volcanic Cascade Range, rises majestically to 10,457 feet. On the western slopes of Lassen Peak, the forest community is dominated by ponderosa pine (*Pinus ponderosa*), white fir (*Abies concolor*), and greenleaf manzanita (*Arctostaphylos patula*), while on the eastern side the principal species are Jeffrey pine (*P. jeffreyi*), lodgepole pine (*P. contorta* ssp. *murrayana*), and rabbitbrush (*Chrysothamnus nauseosus*). At progressively higher elevations forests are dominated by red fir (*Abies magnifica*), mountain hemlock (*Tsuga mertensiana*), and whitebark pine (*P. albicaulis*).

Lassen Volcanic National Park encompasses and protects most of the subalpine and alpine habitats within the region. Though the park includes some mixed coniferous forest at lower elevations, the original boundaries were drawn to exclude the best marketable timber areas.

The National Park Service, which has administered this park since 1916, is mandated to protect its natural features while allowing maximum recreational opportunities—a dual objective that has sometimes resulted in damage to the native flora. Many of the weedy, non-native plants that have become established in the park, for instance, are in the vicinity of the ski runs, which were expanded in the 1980s.

Despite these inevitable conflicts, the park provides an opportunity to appreciate a rich flora that, for the size of the park, is nearly unequaled in California.

Mary Ann and David Showers

Up to 500 people hike the trail to the summit of Mount Lassen, OPPOSITE, *on a typical summer day, some early enough to catch the shadow the mountain casts over the Great Valley at dawn. • Applegate's paintbrush (*Castilleja applegatei*),* TOP, *a hemiparasitic plant, creates bright spots of color among the rocks during summer months. • Pinedrops (*Pterospora andromedea*),* ABOVE, *a saprophytic member of the heather family, lives on decaying matter found on coniferous forest floors. • Foot traffic can damage sensitive alpine species such as golden draba (*Draba aureola*),* LEFT, *which grows only on Lassen Peak and nearby Mount Eddy. It is one of more than twenty Cascade species that reach their southern range limit within the park.*

MODOC PLATEAU

The grays and greens of sagebrush scrub and juniper woodlands on the Modoc Plateau are deceptively monotonous, for growing here is a rich diversity of plants and habitats found nowhere else in the state. This 6,000-square-mile piece of the Great Basin is a semi-arid, volcanic tableland, a southern extension of the Columbia Plateau of eastern Oregon and Washington.

Basalt-capped plateaus formed from vast lava flows extend across the seemingly endless tablelands roughened with escarpments, dormant volcanoes, and valleys. Rimrock Valley, Porcupine Rim, and the Warner Mountains are highlands formed from upthrust blocks. Downthrust blocks contain lakes, extensive meadows, or pluvial lakebeds such as Surprise Valley and the Madeline Plains.

The flora of the Modoc Plateau is unique in California owing to its Great Basin affinities and its location at a crossroads of the Sierran and Cascade

floristic regions. The cold-desert climate is extreme with long months of freezing winter temperatures followed by summer drought and daytime temperatures generally exceeding 90 degrees F. Limited soil moisture during warm months creates a significantly shortened growing season. Sagebrush scrub and juniper woodlands cover the expansive tablelands, while conifer forests top mountains and scattered hills.

Although the region does not support a large number of rare plant species, those that occur on the plateau are generally associated with a specific soil or bedrock. Columbia Plateau endemics, such as Greene's mariposa lily (*Calochortus greenei*), comprise one group of rare plants. Another group of the Modoc Plateau's rare species, such as stoloniferous pussytoes (*Antennaria flagellaris*) and Raven's lomatium (*Lomatium ravenii*), are rare in California but more abundant further east in high-montane outposts of the Great Basin. Recent lava flows provide habitat for

other rarities such as Baker's cypress (*Cupressus bakeri*).

The plateau's vernal pools provide habitat for rare species of the California floristic province, including the tiny Bogg's Lake hedge-hyssop (*Gratiola heterosepala*) and slender Orcutt grass (*Orcuttia tenuis*), which may have migrated east from the Central Valley along the lowland Pit River corridor. Numerous more common species of the Sierra-Cascade foothills have entered the region via this route.

Most of the land on the Modoc Plateau is federally owned. Several unique areas, such as Ash Valley, have been set aside for protection. Coordinated resource management plans are being used to coordinate efforts between federal, state, and local interests to reduce impacts on biological resources.

James D. Jokerst

Ancient lake beds, playas, and vernal pools with poorly drained soils support a mix of Californian and Great Basin species on the Modoc Plateau, FAR LEFT. • *Spring brings great washes of color to the plateau with Basin rayless daisy (Erigeron aphanactis var. aphanactis) and thread-leaved daisy (Erigeron filifolius var. filifolius),* OPPOSITE TOP, *and field owl's-clover (Triphysaria eriantha ssp. eriantha) and Baci's downingia (Downingia bacigalupii),* OPPOSITE MIDDLE. • *When winter snows melt away, wet meadows burst into color with bright yellow members of the carrot family (Lomatium sp.) and purple camas (Camassia quamash),* ABOVE, *once a popular food for Native Americans.* • *Stately Washoe pine (Pinus washoensis),* LEFT, *grows at the crest of the Warner Mountains bordering the Modoc Plateau.*

SAN FRANCISCO BAY REGION

Though only about 6,000 square miles in area, less than four percent of California, the San Francisco Bay region presents an unparalleled diversity of climate, soils, plant communities, and vegetation types and an abundance of native plant species, including many that are extremely rare. Within the region, daily and annual temperature cycles vary tremendously from one microclimate to another; one can be shivering in the dense summer fog along the coast of Marin or San Mateo County while residents of Livermore or Cupertino are suffering temperatures of over 100 degrees F.

Rainfall also varies markedly. San Jose and Antioch receive about thirteen inches of rainfall a year; yet in the Santa Cruz Mountains community of Boulder Creek, only twenty miles from downtown San Jose, about sixty inches of annual rainfall support forests of coast redwood (*Sequoia sempervirens*). Snowfall is rare, but every few years snow closes the roads to the summits of Mount Diablo (3,849 feet) and Mount Hamilton (4,209 feet), and once a decade or so people have been known to ski on Mount Tamalpais (2,610 feet).

The San Francisco Bay region has a complex geologic makeup. In addition to widespread sedimentary rocks and their associated fossils, the region displays rocks of volcanic origin, gran-

ite, serpentine, and limestone. There are also coastal and interior sand dunes formed along the Sacramento River. To botanists, one of the most distinc-

tive substrates here is serpentine, California's state rock, which decomposes to form an infertile nutrient-poor soil that is chemically toxic to plants. Despite this, a number of plant species are able to grow on serpentine, and some of these occur only on these soils. Among them is the Tiburon mariposa lily (*Calochortus tiburonensis*), first discovered in 1973, yet growing within view of the Golden Gate Bridge. Another serpentine endemic is the Presidio manzanita (*Arctostaphylos hookeri* ssp. *ravenii*), now known in the wild as a single shrub on the grounds of San Francisco's Presidio. Its close relative, the Franciscan manzanita (*A. hookeri* ssp. *franciscana*), another serpentine endemic, is extinct in the wild.

With its diverse climate and soils, the Bay region supports an array of vegetation types and plant communities. Coastal regions and some of the interior hills are cloaked with various types of forest and woodland. Plants of the valley grasslands meet and mix with those of the coastal prairies around the bay. Freshwater and saltwater marshes are common, although the extent of the latter has decreased alarmingly since the late nineteenth century.

Vernal pools, with their colorful displays of spring annuals, can still be found in a few areas. Mountain slopes support extensive stands of chaparral and other scrub vegetation. Yet the sand dune vegetation that once covered much of the peninsula now occupied by San Francisco is nearly gone. Urban development here and elsewhere has eliminated much of the plant life. Other communities, especially grass-

lands, have been invaded by weedy exotic plants from other continents. Even the coast redwood forests support their share of such weeds.

Fortunately, the Bay region is blessed with thousands of acres of preserves, parks, and other open space in the public domain. These are home to many hundreds of native plants typical of the region, including most of the rare ones. But the region also is home to nearly six million people, and there are relentless pressures to develop its botanically rich landscapes.

Robert Ornduff

The fabled waters inside the Golden Gate, LEFT, *dominate the landscape of the San Francisco Bay region, an area that encompasses the bay, as well as the rivers and estuaries out to the Sacramento-San Joaquin Delta, and the valleys and mountains that form the nine counties bordering the bay.* • *In the early 1970s, while botanizing on Ring Mountain just across the Golden Gate from San Francisco on the Tiburon Peninsula, Dr. Robert West, physician and amateur naturalist, discovered a flower that didn't fit any existing botanical description. Botanists soon confirmed that it indeed was a new species of mariposa lily growing right next to a well explored, major urban center. The plant was later named Tiburon mariposa lily (Calochortus tiburonensis),* TOP. • *Meadowfoam (Limnanthes alba),* ABOVE, *paints wet meadows and edges of vernal pools a soft white in spring.*

SAN FRANCISCO PRESIDIO

Located at the tip of the San Francisco Peninsula, the San Francisco Presidio overlooks the Golden Gate, where the waters of California's two largest rivers flow out of San Francisco Bay into the Pacific Ocean. The Presidio was the first place in the Bay region to be settled by Europeans, and for more than 200 years it served as an army base. In 1994 it became part of the Golden Gate National Recreation Area, administered by the U.S. National Park Service.

Most of what remains of the city of San Francisco's native plant populations grows on Presidio land, including a high concentration of rare, endemic, and unusual plants. The native plants there are found today in small and extremely vulnerable locations on the undeveloped half of the Presidio's 1,400 acres. Several rare plants remain on the Presidio, including Presidio

clarkia (*Clarkia franciscana*), Marin dwarf flax (*Hesperolinon congestum*), and San Francisco lessingia (*Lessingia germanorum* var. *germanorum*).

Some of the earliest scientific collections of California plants and animals were made here by early explorers. The Army's earlier extensive program of planting non-native trees such as Monterey pine (*Pinus radiata*), Monterey cypress (*Cupressus macrocarpa*), and Australian eucalyptus have shaded out many native plants, and the spread of weedy non-natives such as iceplant (*Carpobrotus* spp.) has crowded out others.

With the change in status from a military base to a national recreation area, and with the increased foot traffic this change will bring, the challenge for the National Park Service will be to manage the area to preserve the remnants of a once diverse and still important coastal flora.

Jacob Sigg

EARLY BOTANISTS ABOARD RUSSIAN EXPLORER SHIPS

In October 1816, French-born Adelbert Chamisso and Johann Friedrich Eschscholtz, born in the Baltic provinces of Russia, visited the Presidio of San Francisco, then under the Spanish flag. They came as naturalist-interpreter and surgeon-naturalist aboard the *Rurik*, a two-masted, square-rigged Russian vessel under the command of Otto von Kotzebue.

During the month of October the two naturalists collected eighty-two species of California plants in or near the Presidio, forty of which were new to European scientists. Of the eighty-two plants Chamisso and Eschscholtz collected from the Presidio, fifty-four species still grow there today.

There are several other common plants in the California flora that were first described from the Presidio, including: the well known wax-myrtle (*Myrica californica*), yerba buena (*Satureja douglasii*), the fragrant California sagebrush (*Artemisia californica*), spring-blooming California lilac (*Ceanothus thyrsiflorus*),

Today, the best known plant collected by botanical explorers Chamisso and Eschscholtz on the Russian Romanzoff Expedition is the California state flower, the California poppy (Eschscholzia californica). The "type" specimen of the poppy (the specimen with which the name is permanently affiliated) still remains in Russia in the St. Petersburg Herbarium.

seaside woolly sunflower (*Eriophyllum staechadifolium*), and California coffeeberry (*Rhamnus californica*).

Ida Geary

The lack of urban development on most of the Presidio's serpentine outcrops has allowed many rare serpentine plants to survive, including the last known individual of Presidio manzanita (Arctostaphylos hookeri ssp. ravenii), OPPOSITE, TOP. • For years Presidio clarkia (Clarkia franciscana), OPPOSITE, BOTTOM, was known only from the Presidio serpentine, but recently this annual plant has been found on serpentine outcrops in the hills above Oakland. • San Francisco wallflower (Erysimum franciscanum), TOP LEFT, once a common bright spot along San Francisco's dunes and coastal scrub, is now much rarer due to habitat losses. • The Presidio, LEFT, was once home to many more native plants than today. Steep cliffside natural gardens pruned by wind and watered by fog were of exceptional beauty. Ocean winds blew sand onshore, creating a vast dune system since covered by roads, buildings, or non-native vegetation.

MOUNT TAMALPAIS

Mount Tamalpais rises to an elevation of 2,610 feet in southern Marin County, making it the tallest and most prominent feature in this part of the San Francisco Bay region. After hiking its many trails through a mosaic of plant communities, one is not surprised to learn that it is one of the richest sites of endemic species in California. Mount Tamalpais is home to seventy-three species of plants found nowhere else in the county. More than fifty species reach their southern range limits on the mountain, and another twelve to fifteen reach their northern limits here.

These restricted and sometimes rare plants of Mount Tamalpais are diverse in origin. Some, such as serpentine

reed grass (*Calamagrostis ophitidis*) and the early blooming coastal shrub western leatherwood (*Dirca occidentalis*), are paleoendemics representing ancient lineages. Others, such as the shrubs Mount Tamalpais manzanita (*Arctostaphylos hookeri* ssp. *montana*) and musk brush (*Ceanothus jepsonii* var. *jepsonii*), and many beautiful annual wildflowers such as the Tamalpais jewelflower (*Streptanthus batrachopus*), have evolved more recently and are often found in serpentine areas.

Many features of Mount Tamalpais account for this impressive display of plant endemism. Soils are developed from ancient Franciscan rock or from serpentine outcrops which support serpentine grasslands, chaparral, and several forest types. The north slope of the mountain is one of the wettest places in the Bay Area, receiving an average of fifty-two inches of rain annually. Coastal summer fog wraps around the slopes of the mountain enclosing redwood forests and such restricted communities as maritime chaparral with rare species like Mason's ceanothus (*Ceanothus masonii*), glory brush (*C. gloriosus* var. *exaltatus*), and Marin manzanita (*Arctostaphylos virgata*).

Most of the Mount Tamalpais area is in public ownership, principally the Golden Gate National Recreation Area, Mount Tamalpais State Park, Muir Woods National Monument, Marin Municipal Water District, and Marin County Open Space District. These agencies are beginning to manage vegetation and rare species on Mount Tamalpais using an ecosystem approach to preserve the diverse plant life for future generations to enjoy.

V. Thomas Parker

Mount Tamalpais, LEFT, *is one of the most popular natural areas in the United States, drawing many thousands of tourists and local residents who enjoy its abundant natural treasures.* • *In early spring, checker lily* (Fritillaria affinis *var.* affinis), OPPOSITE, TOP, *is quite common on brushy and wooded slopes.* • *Baby blue-eyes* (Nemophila menziesii) (THIS PAGE, CLOCKWISE FROM TOP), *common on grassy or brushy slopes, occurs in variable forms with colors ranging from vivid blue to white.* • *Calypso orchid* (Calypso bulbosa), *at the southern end of its range on Mount Tamalpais, is sometimes decimated by feral pig rootings.* • *False indigo* (Amorpha californica) *is one of several shrubs on the mountain that stump sprout following a fire.* • *Eight species, among them the bright red-flowered Mount Tamalpais thistle* (Cirsium hydrophyllum *var.* vaseyi) *and the Tamalpais jewelflower* (Streptanthus batrachopus), NOT PICTURED, *are found nowhere else in the world.* • *Bush poppy* (Dendromecon rigida) *is a characteristic member of the chaparral community on the south side of the mountain.*

POINT REYES PENINSULA

Millions of years ago, a piece of the Pacific plate now called the Point Reyes Peninsula first struck the North American plate at a point about 300 miles south of its present location. Movement along the San Andreas fault has caused it to inch northward, so that today the peninsula lies about thirty miles north of San Francisco, separated from the mainland by the San Andreas rift zone underlying Tomales Bay, the Olema Valley, and Bolinas Lagoon.

Because of the peninsula's unique geological ancestry, its soils and plant communities contrast markedly with those of the mainland only a few hundred yards away. Granite, overlain in places by marine sediments, forms Point Reyes, while mainland rocks east of the fault are largely from an ancient Franciscan formation.

Inverness Ridge is the granitic backbone of the peninsula. To the south the ridge is forested by Douglas-fir; to the north, where granite is more exposed, by a relictual stand of Bishop pines, once a more widespread species. Notably missing on the peninsula are redwood forests and serpentine outcrops, which occur only east of the rift zone.

The ridges, rugged bluffs, a ten-mile stretch of beach, and many fingered inlets provide a remarkable di-versity of habitats: sand dunes, coastal bluffs, salt marshes, freshwater marshes, sag ponds, forests, and grasslands. Plant diversity is further enriched by the peninsula's climate, which is strongly influenced by the sea and by the persistent summer fogs differentiating it from California's generally Mediterranean climate. More than 850 native and introduced plant species, including thirty-two rare species, grow on the peninsula.

Most of the Point Reyes Peninsula is included in the Point Reyes Na-

tional Seashore, which provides cru-cial protection. Part of the peninsula's central grasslands is by law leased to beef and dairy ranchers, but the National Seashore regulates grazing and monitors natural resources in this pas-

toral zone. Each year, dedicated volunteers from the California Native
Plant Society survey the health and
extent of the peninsula's rare plant
populations.

Virginia Norris

Point Reyes Peninsula, ABOVE, *is famous for its fields of spring wildflowers on windswept ocean bluffs.* • *The columnar stamen tube of checkerbloom* (Sidalcea malvaeflora), OPPOSITE, TOP, *clearly identifies this spring beauty with the mallow family.* • *Footsteps of spring* (Sanicula arctopoides), OPPOSITE, MIDDLE, *is an early harbinger of spring on ocean-facing bluffs.* • *The world's only remaining population of the bristly Sonoma spineflower* (Chorizanthe valida), OPPOSITE, BOTTOM, *thought to be extinct for 77 years and rediscovered in 1980, is found here on grazed coastal prairie.*

VINE HILL BARRENS

The native vegetation that for thousands of years survived in the nutrient-poor, highly acidic, weathered sandstone soils of Sonoma County's Vine Hill barrens is virtually gone today, replaced by orchards and vineyards. All that is preserved of the native species grows on a 1.5-acre preserve located between Sebastopol and Santa Rosa. This preserve, the Vine Hill Preserve, is owned by the Milo Baker Chapter of the California Native Plant Society.

Despite the dramatic losses, the unusual and rare plants endemic to Vine Hill, particularly the manzanitas and California lilacs, have attained fame in horticultural circles. Perhaps most noteworthy is the Vine Hill manzanita (*Arctostaphylos densiflora*), which grows in low, sprawling mats festooned with clusters of white to pinkish bell-shaped flowers typical of manzanitas. Numerous natural hybrids of various manzanita species occur at Vine Hill and are of special interest to botanists. Vine Hill manzanita, Baker manzanita (*Arctostaphylos bakeri*), and the beautiful common manzanita (*A. manzanita*) all are found at the preserve and along Guerneville Road and all hybridize. The endemic Vine Hill California lilac (*Ceanothus foliosus* var. *vineatus*), nearly extirpated from Sonoma County, is now known in the wild only from Vine Hill. Two other plants, more typically found along the immediate coast where summer fog is reliable, are beargrass (*Xerophyllum tenax*) and salal (*Gaultheria shallon*).

Glenn Keator and Philip Van Solen

The Vine Hill Preserve, TOP, provides an example of a barren, a formerly more extensive native plant community where plants have developed special adaptations to grow in nutrient-poor soils. • The attractive and endangered Vine Hill manzanita (Arctostaphylos densiflora), BELOW LEFT, is known to gardeners as the cultivar 'Howard McMinn'. • Botanists have attempted to reestablish the endangered Vine Hill clarkia (Clarkia imbricata), BELOW, on the preserve; however, the results to date are not promising. Although easy to cultivate for garden use, this lovely godetia-type clarkia is currently known from only one natural population in the wild.

THE ENTWINED LIVES OF PLANTS AND INSECTS

The lives of plants and insects are linked in intricate and intimate ways. There are thousands of species of beetles, bees, wasps, flies, moths, and butterflies that pollinate flowers. Some beetles are of special interest because they were the earliest flower-visiting insects and, over many eons, have evolved unique relationships with other organisms. For example, the common checkered beetle (*Trichodes ornatus*), in the family Cleridae, has developed a flower dependency with a complexity found in no other insects.

This beetle is widespread in California from sea level to 10,000 feet. Its larvae hatch from eggs laid in a flower, preferably common plants such as yarrow (*Achillea millefolium*) or checkerbloom (*Sidalcea malvaeflora*). The larvae crawl aboard a flower-visiting bee or wasp, which unknowingly transports them back to their own nests. Once in a nest cell of the carrier, the checkered beetle larva waits until the carrier's larvae are fully grown and then quickly consumes them. By April or May, after overwintering and pupating, adult checkered beetles emerge to spend their entire lives in flowers. Here, they feed on pollen, nectar, other insects, and even their own mates. The female then lays her eggs in a flower where emerging larvae will again grab a ride to another species' nest for its food supply.

Through all of this strange behavior, the checkered beetle performs a vital function as floral pollinator, critically important as it ensures that there will be flowers in the future for its descendants. To increase their effectiveness as pollinators, many flower-frequenting beetles have an abundance of pollen-collecting hairs on their forebodies.

Daytime exposure to predators in a flower can be hazardous to an insect or spider, and thus some species are protected by camouflage or, in the case of the checkered beetle, by conspicuous yellow and black banding simulating that of a stinging wasp or bee. The bee association has long been assumed, perhaps even by Aristotle. In naming the checkered beetle, *Trichodes apiarus*, the eighteenth-century father of taxonomy, Carl Linnaeus, chose the name for honeybee (*Apis*) for the beetle's species name.

Edward S. Ross

A typical native leaf cutter bee, shown on gumplant (Grindelia stricta), ABOVE, *carries a pollen load on its "belly," unlike most other bees which carry pollen on leg hairs. The white gum of the bud under the open yellow flower inhibits seed predator beetles from laying eggs in gum plant seeds.* • *The common checkered beetle* (Trichodes ornatus), LEFT, *has developed a flower dependency with a complexity found in no other insects. Here, the pollen-covered beetle seeks nectar in checkerbloom* (Sidalcea malvaeflora). *Its yellow and black banding mimics the banding pattern marking yellow jackets, a species that birds avoid.*

SANTA ROSA PLAIN

In the undulating plains between the rapidly growing cities of Santa Rosa and Windsor, one can still chance upon the breathtaking springtime beauty of Sonoma County's vernal pools. Though less famous than those of the Great Valley and Southern California, the pools on the Santa Rosa Plain are well known to local botanists. And they are home to rare and endangered plant species found growing nowhere else.

These vernal pools occur on flat or hummocky terrain underlain by an impervious claypan substrate that ensures a water table close to the surface. Interspersed with valley oak woodlands, most of the vernal pools occur on poorly drained acidic clay loam across the broad valley floor east of the Laguna de Santa Rosa, extending east toward the foothills of the inner North Coast Range and north to Windsor. As the cities of Santa Rosa and Windsor

have expanded, many of the vernal pools have disappeared. Those remaining, together with their rare plant populations, are threatened by continuing urban development, changes in agricultural uses, and summer irrigation.

The U.S. Fish and Wildlife Service is funding the development of a vernal pool ecosystem preservation plan for the Santa Rosa Plain. The general goals of the plan are to preserve the diverse plants and animals that live in the Santa Rosa Plain vernal pool ecosystems and their related watersheds; to develop methods to resolve conflicts between landowner, agency, and conservation interests in an effective and timely manner; and to simplify the permitting process for development. Through an essential process beginning with public participation, federal, state, and local agencies as well as agricultural, resource conservation, development, and landowner interests are now cooperating to achieve permanent protection of the Santa Rosa Plain vernal pools.

Betty L. Guggolz

While some vernal pools produce spectacular concentric rings of flowers, most pools form in interconnected swales, TOP, *in wet years a lovely mosaic of green foliage and gold from the rare Burke's goldfields* (Lasthenia burkei). • *During early spring,* LEFT, *bright yellow rings of the endangered Sonoma sunshine* (Blennosperma bakeri) *and soft white pools of Sebastopol meadowfoam* (Limnanthes vinculans) *can be seen on the Santa Rosa Plain.* • *The beauty of this flower,* OPPOSITE, TOP, *is enhanced by the creamy white tips of the yellow rays giving the plant its common name, tidy-tips* (Layia fremontii). • *Plants growing in vernal pools are primarily annual species such as the endangered soft white Sebastopol meadow-foam,* ABOVE, *and Burke's goldfields* (Lasthenia burkei), RIGHT.

MOUNT DIABLO

From the 3,849-foot summit of Mount Diablo, one can see farther in all directions than from almost any other point on earth, from Mount Lassen to Half Dome to the Farallon Islands. Located only thirty miles northeast of San Francisco, Mount Diablo is an isolated landmark peak of the Central Coast Range with a complexity of soils and geology hosting a rich array of over 500 species of native vascular plants. Its peak averages twenty-two inches of annual rainfall, while warmer lower slopes receive several inches less.

For plants adapted to higher elevations, the surrounding low-lying valleys form a barrier to migration. Coulter pine (*Pinus coulteri*), for example, finds its northern limit on Mount

Diablo, and the beautiful common manzanita (*Arctostaphylos manzanita*) and western viburnum (*Viburnum ellipticum*) occur at the southern limit of their ranges here. In the wooded canyons of the lower slopes and on brushy hillsides of the north side of the mountain, the delicate whitish flower clusters of the hop tree (*Ptelea crenulata*), a plant restricted to the Inner Coast Ranges, add fragrance to the air in

early spring. The upper talus slopes and outcrops of Mount Diablo are good places to observe some of the rarest endemic plants on the mountain such as Mt. Diablo fairy lantern (*Calochortus pulchellus*), growing only on Mount Diablo, rock sanicle (*Sanicula saxatilis*), Mount Diablo jewelflower (*Streptanthus hispidus*), and the fiddleneck blooms of Mount Diablo phacelia (*Phacelia phacelioides*).

Much of Mount Diablo is protected by state, regional, and local parks, namely Mount Diablo State Park and

Spring brings a rosy hue to the blue oaks (Quercus douglasii), RIGHT, *covering the slopes of Mount Diablo.* • *The Mount Diablo fairy lantern* (Calochortus pulchellus), ABOVE, *was a favorite of Willis Linn Jepson, author of the* Manual of Flowering Plants of California, *first published in 1925, and was chosen by him as his logo. It is retained as the logo of the Jepson Herbarium at the University of California, Berkeley.* • *Early morning fog softens the slopes near Curry Point,* OPPOSITE, TOP.

Diablo Foothills Regional Park, as well as Lime and Shell Ridges and Sugarloaf Recreation Area, which are owned by the cities of Walnut Creek and Concord. Hikers can enjoy spring wildflowers and scenic views from the many public trails on Mount Diablo. Major acquisitions of privately held lands, especially on the north, east, and west sides, are still needed to protect important plant habitat as well as to increase open space for residents of the Bay Area.

Mary L. Bowerman and Susan D'Alcamo

LARGE-FLOWERED FIDDLENECK

Presumed extinct for twenty years, large-flowered fiddleneck (*Amsinckia grandiflora*) was rediscovered in 1938 on the steep slope of a ravine in western San Joaquin County. Though it once ranged from the Mount Diablo foothills to the Inner Coast Ranges near Livermore, this distinctive annual of the borage family was never common.

It once again grows near Stewartville, a coal-mine ghost town south of Antioch in Contra Costa County as the result of an active recovery program initiated in 1987 by the California Department of Fish and Game. One of the first of such programs for an endangered plant, it has been carried out by researchers at Mills College through contributions to the Endangered Species Tax Check-Off Fund, as well as through funding from the U.S. Fish and Wildlife Service.

Ann Howald

Large-flowered fiddleneck (Amsinckia grandiflora) *disappeared from most of its range because of ecological disruption, probably a combination of competition from introduced annual grasses and alterations of fire and grazing regimes.*

ANTIOCH DUNES

The native vegetation at Antioch Dunes in eastern Contra Costa County is reminiscent of arid Southern California. Deep patches of bone-white sand support low shrubs otherwise found in distant coastal sage scrub, chaparral, and desert habitats.

Why should this desert-like sandy habitat occur in Contra Costa County? It was created over centuries by the rain-swollen Sacramento and San Joaquin rivers, which would often break loose from their channels in spring, flooding great tracts of valley lowland. The swift currents carried tons of new sediment, which spread across the floodplain. As the rivers receded into their summer channels, winds blew the riverside sands into low dunes and sand fields.

Unfortunately, non-native weedy plants have invaded the dune vegetation, aided by human disturbance and the lack of new river sand. A thick sward of exotic weeds such as ripgut grass (*Bromus diandrus*), wild oats (*Avena fatua*), and Russian thistle (*Salsola tragus*) now prevents the natural processes of dune growth and movement.

Habitat destruction, combined with habitat alteration by non-natives, has put two beautiful plants at risk of extinction: the Contra Costa wallflower

(*Erysimum capitatum* var. *angustatum*) and the Antioch Dunes evening primrose (*Oenothera deltoides* ssp. *howellii*). Neither can maintain large stable populations for lack of suitable dune habitat.

On their adjacent properties, both Pacific Gas & Electric and the U.S. Fish and Wildlife Service have experimentally created and restored dunes as habitat for the endemic plants and the endangered Lange's metalmark but-

terfly. Using more than 7,000 cubic yards of river sand donated by PG&E, the Service has built two and one-half acres of new sand dunes and planted them with the two rare species. Although this effort requires constant maintenance and monitoring, it has increased our understanding of how to restore degraded ecosystems and the endangered species they support.

Bruce M. Pavlik

The Antioch Dunes, LEFT, near the town of Antioch, are remnants of this once wild and mighty river system. Despite the cessation of annual flooding and after years of sand mining, relicts of this dune system still support a singular assemblage of plants and animals, some of which are found nowhere else in the world. • The striking Antioch Dunes evening primrose (Oenothera deltoides ssp. howellii), ABOVE, is known only from a small population on the Antioch Dunes in the Sacramento-San Joaquin Delta. It is unlikely that this evening primrose was ever common.

SAN BRUNO MOUNTAIN

Given its proximity to San Francisco and the suburbs of the Peninsula, it is remarkable that San Bruno Mountain has managed to remain in a relatively pristine natural state until as recently as thirty years ago. Today, however, the mountain's remaining acres of undeveloped land are encroached upon by four expanding cities—Brisbane, Daly City, Colma, and South San Francisco—and a quarry, industrial developments, a landfill, and radio and television transmission towers. A relentless encroachment of weedy non-native plants also threatens the integrity of the mountain's remaining plant communities.

This mountain still holds many natural treasures. The topography and soils of San Bruno Mountain, which is linked geologically and biologically to the Santa Cruz Mountains, are sufficiently varied to support several plant communities: grassland, shrubland, woodland, wetland, and even a dune scrub area on the northwest flank, several miles from the ocean. At least 659 species of plants have been found here, of which about one-third are non-natives.

Owned by the Crocker Estate from 1884, the mountain remained largely undisturbed until the 1960s, when the Crocker Land Company undertook plans for developing much of it. Citizen-based conservation groups became especially active when the mission blue butterfly, a federally listed endangered species, was found on the mountain. Ultimately, a 3,000-acre San Bruno Mountain State and County Park was established, protecting the butterfly's habitat and many of the rare plants.

In 1982, San Mateo County adopted a habitat conservation plan for the butterfly, the first in the nation. But a decade later, controversy about its effectiveness remains. Massive invasions of introduced weeds continue to threaten both the rare plants and the butterfly. Each year, as weeds broaden their hold on the mountain, eradication costs increase, and rare and other native species are at further risk. Continuing public support, volunteer efforts, and local agency attention will be needed to curb and reverse the steady march of introduced weeds across this natural treasure.

Elizabeth McClintock

Tens of thousands of freeway commuters speed by San Bruno Mountain's gently rolling flanks every day, with scarcely a glance at this tarnished jewel, BELOW. • *San Bruno Mountain manzanita* (Arctostaphylos imbricata), ABOVE, *is known from only five populations there and is threatened by changes in fire regimes and urbanization.*

EDGEWOOD COUNTY PARK

In 1993, following an outpouring of public sentiment, the San Mateo County Board of Supervisors abandoned plans to build a golf course and instead designated Edgewood County Park as a natural preserve. The conservation of an important wildflower area for future generations was a critical decision. Because of Edgewood County Park's exceptional biological diversity, the California Department of Fish and Game recognizes the park as a significant natural area in California.

A wealth of botanical riches has survived years of rampant off-road vehicle activity, vandalism, and trash dumping on the 467-acre area known as Edgewood County Park. This small park provides habitat for 440 plant species, eleven of which are rare, threatened, or endangered.

Nine distinct plant communities occur in the park: grassland, serpentine bunchgrass grassland, chaparral, mixed serpentine chaparral, foothill woodland, mixed-evergreen forest, redwood forest, riparian, and wetlands. Soils developed from a Franciscan formation (serpentine, greenstone, and sandstone) and the Santa Clara formation (sands, gravels, and clays) became intermixed during past geologic activity and created a specialized habitat.

Two plant communities, serpentine bunchgrass grassland and mixed serpentine chaparral, are considered threatened because of the extensive development in the Bay Area that eliminates these habitats. The Santa Clara Valley chapter of the California Native Plant Society offers wildflower hikes for the public at Edgewood County Park every spring.

Susan Sommers

Edgewood Park, ABOVE LEFT, *lying above a fog belt that forms along the San Andreas faultline and in the rain shadow of the Santa Cruz Mountains to the west, becomes carpeted with wildflowers in early spring.* • Fountain thistle *(Cirsium fontinale* var. *fontinale),* FAR LEFT, *is known from only four populations in the vicinity of the park and Crystal Springs Reservoir, where it grows in specialized serpentine seeps.* • California plantain *(Plantago erecta),* LEFT, *growing in a serpentine grassland is a food plant for the bay checkerspot butterfly.* • *Among the very rare plants found here, the small aromatic San Mateo thornmint (Acanthomintha duttonii),* TOP RIGHT, *is one of the rarest, the victim of San Mateo County's extensive urbanization.* •San Mateo woolly sunflower *(Eriophyllum latilobum),* ABOVE, *was rediscovered in 1981 and is known from only one population in the world.* • *Three endangered arthropod species, the bay checkerspot butterfly,* BELOW, *and two harvestman spiders, are restricted to serpentine soils, the latter only to those in Edgewood Park.*

SANTA CLARA VALLEY

The Santa Clara Valley, a fertile, low-elevation valley drained by Coyote Creek, is bordered by the Santa Cruz Mountains on the west and the Mount Hamilton Range on the east, and represents a typical valley of the Coast Ranges. Like much of coastal California, its geology is complex; serpentine seeps and Franciscan melange soil types support several noteworthy plants.

Today, vegetation in the Santa Clara Valley is mostly native perennial bunchgrass or non-native annual grassland, occurring as grazed pasture along the foothill slopes. Riparian woodland dominated by sycamore (*Platanus racemosa*), willow (*Salix* spp.), and mule fat (*Baccharis salicifolia*) occurs along Coyote Creek, while coastal sage scrub, mixed-serpentine chaparral, and northern mixed-chaparral communities are still found in patches on the lower slopes and, more broadly, on the upper slopes of the hills surrounding the valley. Coast live oak woodland occurs in shallow, narrow canyons.

Agriculture and grazing modified the Santa Clara Valley early on, followed by urban development, which continues to be the greatest threat to native plants. The rare Marin dwarf flax (*Hesperolinon congestum*) and Contra Costa goldfields (*Lasthenia conjugens*) once grew in vernal pools in the valley, but both are believed to have been eliminated when the pools were drained for agriculture.

As urbanization continues to expand into the Santa Clara Valley, more of its plant species are becoming threatened. Conservationists are increasingly concerned about the serpentine grasslands because they support the greatest diversity of native plants and the majority of rare and endemic plants remaining in the Santa Clara Valley. No areas have been set aside to protect the special habitat of these species.

Niall McCarten

The handsome valley oak (Quercus lobata), OPPOSITE, *once a common sight in the Santa Clara Valley, is becoming increasingly rare. Mature trees were cut for firewood or pasture management; seedlings are eaten by deer, cattle, and rodents.* • *Early-blooming fragrant fritillary* (Fritillaria liliacea), ABOVE LEFT, *is threatened by grazing and loss of habitat to agriculture and urbanization.* • *Cattle introduced by early settlers brought European annual grasses, which replaced native bunchgrasses. The introduced wild oat* (Avena fatua) *is shown,* ABOVE RIGHT, *growing under an old coast live oak.* • *The upper two petals of johnny-jump-ups* (Viola pedunculata), RIGHT, *are a dark red-brown on the outside.*

CENTRAL COAST AND SOUTH COAST RANGES

California's spectacular Central Coast and South Coast ranges and their related dunes, terraces, plains, and valleys provide habitat to roughly one-third of all the native plant species that occur in the state—about 2,200 in all. Of those, about 175 are endemic, occurring nowhere else in the world.

The Central and South Coast ranges extend southward from San Francisco almost 200 miles to meet the east-west trending Santa Ynez Mountains, just north of Santa Barbara in the Transverse Range, and from the coast inland about fifty miles to the western edge of the San Joaquin Valley. The more northern Santa Lucia Range rises to almost 6,000 feet. These coastal mountains are interrupted by major bays and coastal plains narrowing into valleys at Santa Cruz, Monterey Bay, Morro Bay, and the Santa Maria River Basin. Between the various ranges that comprise the Central Coast Ranges are several valleys, the largest of which are the Pajaro and Salinas river valleys.

Major coastal dune systems occur at Monterey, Morro Bay, and the mouth of the Santa Maria River, a product of the erosion and deposition cycle found in coastal watersheds. A specialized beach flora, well adapted to strong ocean winds and salt spray, has developed on the dunes. Remnants of uplifted ancient dune systems are found at Bonny Doon and Fort Ord in Monterey County, Morro Bay, and the Nipomo Mesa in San Luis Obispo County, and in the vicinity of Burton Mesa in Santa Barbara County. These ancient remnant sites support unique floras with many rare species.

Oak woodland and coastal scrub are the dominant plant communities in the central coastal area. Coast red-

wood (*Sequoia sempervirens*) grows in stately groves near Santa Cruz and in coastal canyons as far south as Salmon Creek Canyon in the Santa Lucia Range of Monterey County. Both the San Rafael and Santa Lucia ranges support montane coniferous forests at higher elevations.

A few remnants of grasslands that once covered the valleys of the central coast remain; they have largely been displaced by agriculture and urbanization. Most are now dominated by introduced annual grasses. By contrast, some of the coastal terraces and prairies are still dominated by a native perennial oatgrass (*Danthonia californica*), which forms lawn-like stands, and in spring by colorful wildflower displays.

Increasingly, native plant habitat is being lost to urbanization, brushland conversion, dredging, mining, off-road vehicle activities, and non-native plant invasions. Exotic French broom (*Genista monspessulana*), which is spreading at an alarming rate, increasingly threatens grassland remnants. Pampas grass (*Cortaderia jubata*) is invading the coastal sage scrub along coastal cliffs and slopes, degrading one of California's classic plant communities.

Malcolm McLeod

From Monterey south, the Central Coast Range rises directly out of the sea and generally parallels the coast to the southern boundary of San Luis Obispo County. Stands of giant coreopsis (Coreopsis gigantea), BELOW, grow with stout fleshy stems and brilliant yellow daisy-like flowers on the sea terrace at Point Arguello, and on the Nipomo Dunes, farther south along the ocean. • A white form of sticky monkeyflower (Mimulus aurantiacus), BELOW LEFT, grows on rocky hillsides and edges of chaparral in the Monterey area. • Prickly phlox (Leptodactylon californicum), LEFT, opens its flowers by day and adds a bright note to the coastal strand and forest edge.

BONNY DOON SANDHILLS

The Bonny Doon and nearby Ben Lomond sandhills of Santa Cruz County, separated by the San Lorenzo River, are islands of uplifted, ancient marine sand deposits. They are found four to ten miles inland, and some attain as much as 800 feet in elevation. Although up to sixty inches of annual rain falls here, the sandhills appear much drier than the surrounding forests because of their well drained sandy soils. Characterized by many as "biological islands," the nutrient-poor sandhills support a rare association of species found nowhere else in the world. The endangered Santa Cruz wallflower (*Erysimum teretifolium*) and Ben Lomond spineflower (*Chorizanthe pungens* var. *hartwegiana*) brighten sandstone beds under the pines.

Never common, the sandhill communities have been greatly diminished. Sand quarries have mined the centers of the best-developed deposits, and residential development has taken a toll around the fringes.

Several acres of sand parkland, maritime chaparral, and ponderosa pine habitat have been preserved at Quail Hollow Ranch County Park. Mining interests, environmental organizations, and local residents have begun working together in this area to preserve the best remaining high-quality example of sand parkland. The Bonny Doon Ecological Reserve, purchased in 1989 and 1990 by the California Department of Fish and Game with the assistance of The Nature Conservancy, protects 600 acres of dense ponderosa pine and maritime chaparral, including one of five known stands of Santa Cruz cypress.

Stephen McCabe and Deborah Hillyard

The most distinctive and biologically diverse community within the sandhills is known locally as "sand parkland," LEFT, *because of the widely spaced park-like stands of ponderosa pine (Pinus ponderosa) on white infertile soils. Sandstone beds, often rich in fossil sand dollars, are found under the pines along with an ephemeral flora of spring-blooming annual and perennial plants.*
• *The endemic Bonny Doon manzanita (Arctostaphylos silvicola), in fruit,* ABOVE, *and lichen-covered,* BELOW, *dominates the maritime chaparral habitat of the sandhills.*

FORT ORD

For generations, hundreds of thousands of servicemen have honed their war skills on the 28,000 acres of the Fort Ord Army base on the southern shore of Monterey Bay. Despite this heavy use, 20,000 of those acres still provide relatively undisturbed examples of seven different native plant communities.

Adjacent to the bay, four miles of coastal dunes provide habitat for several rare dune species, including sand gilia (*Gilia tenuifolia* var. *arenaria*), Monterey ceanothus (*Ceanothus cuneatus* var. *rigidus*), Eastwood's goldenbush (*Ericameria fasciculata*), coast wallflower (*Erysimum ammophilum*), and Monterey spineflower (*Chorizanthe pungens* var. *pungens*). Coastal scrub gradually grades into maritime chaparral farther inland. Fort Ord boasts one of the largest high-quality stands of maritime chaparral in California. Two forms of chaparral are found here: sandhill maritime and Aromas Formation maritime chaparral, each characterized by different species of manzanita. In the base's interior are coast live oak woodlands, riparian pockets, and native grasslands with scattered vernal pools.

Fort Ord serves as an example of the important opportunity sound planning provides to conserve the great biological diversity present on many of California's military bases scheduled for closure. Government agencies and private organizations, such as the California Native Plant Society, have worked together to afford special protection to Fort Ord's wildlands. Because of its significant natural values, much of the former base will be preserved and managed as habitat and open space by the U.S. Bureau of Land Management, the California Department of Parks and Recreation, and the University of California Natural Reserve System.

Mary Ann Matthews and Nicole Nedeff

Like other military bases near urban areas, Fort Ord has served as a refuge for rare and endemic species, LEFT, *that have virtually disappeared from surrounding developed areas. • The rare Monterey spineflower (Chorizanthe pungens var. pungens) and the bright, somewhat more common Indian paintbrush (Castilleja latifolia),* TOP, *grow side by side, well adapted to life in the dunes. • The rare Monterey ceanothus (Ceanothus cuneatus var. rigidus),* ABOVE, *occurs on ancient relict dunes from the Pleistocene epoch. It blooms in February and March.*

MONTEREY BAY DUNES

onterey Bay's sand dunes hug the shore from the Salinas River south nearly to the Carmel River, buffering the region's inland terraces from the Pacific Ocean. Although these dunes are geologically young, human activities are reducing the extent of the dune system, lowering the dune profile, degrading the native vegetation, and accelerating the invasion of exotic plants.

Native coastal dune scrub, a community of low-growing annual and perennial flowering plants and short-lived shrubs, is the typical vegetation of the Monterey Bay dunes. While many dune scrub species are widespread along the Pacific Coast, several are found only in the Monterey Bay dunes and have become endangered as the dune system has declined.

Dunes and dune plants are fragile and easily damaged. The Monterey Bay dunes have been eroded and reduced by off-road vehicles, sand mining, and urbanization. Today, most of the remaining dune system occurs in a thin band west of the Coast Highway. Much of this area is managed by the California Department of Parks and Recreation, which has initiated a stewardship program that includes conserving the native dune ecosystem by restoring degraded habitats, reestablishing endangered plants, and increasing public awareness of the value of California's irreplaceable coastal dunes.

Deborah Hillyard

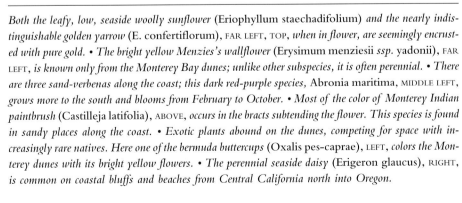

Both the leafy, low, seaside woolly sunflower (Eriophyllum staechadifolium) *and the nearly indistinguishable golden yarrow* (E. confertiflorum), FAR LEFT, TOP, *when in flower, are seemingly encrusted with pure gold.* • *The bright yellow Menzies's wallflower* (Erysimum menziesii *ssp.* yadonii), FAR LEFT, *is known only from the Monterey Bay dunes; unlike other subspecies, it is often perennial.* • *There are three sand-verbenas along the coast; this dark red-purple species,* Abronia maritima, MIDDLE LEFT, *grows more to the south and blooms from February to October.* • *Most of the color of Monterey Indian paintbrush* (Castilleja latifolia), ABOVE, *occurs in the bracts subtending the flower. This species is found in sandy places along the coast.* • *Exotic plants abound on the dunes, competing for space with increasingly rare natives. Here one of the bermuda buttercups* (Oxalis pes-caprae), LEFT, *colors the Monterey dunes with its bright yellow flowers.* • *The perennial seaside daisy* (Erigeron glaucus), RIGHT, *is common on coastal bluffs and beaches from Central California north into Oregon.*

MONTEREY PINE-CYPRESS FOREST

More widespread during the cool, moist Pliocene epoch, today Monterey pine (*Pinus radiata*), Monterey cypress (*Cupressus macrocarpa*), and Gowen's cypress (*C. goveniana*) are restricted to scattered locations along the central coast of California.

Native Monterey pine forests exist in only three mainland locations: Cambria, the Monterey Peninsula, and Swanton near Año Nuevo, all characterized by persistent summer fog and nutrient-poor soils. Two small populations, genetically distinct from mainland stands, occur on Cedros and Guadalupe islands off Baja California. Worldwide, however, Monterey pine is grown commercially in a number of countries, notably, Chile, New Zealand, and Australia, where it is the number-one forest plantation species.

Natural groves of Monterey cypress are confined to only two localities: a narrow coastal strip of windswept granitic headlands at Pebble Beach and across Carmel Bay at Point Lobos. Gowen's cypress is restricted to two natural groves, one on Huckleberry Hill in the Del Monte Forest of Pebble Beach, just inland from the Monterey cypress forest, and the other inland from Point Lobos. Huckleberry Hill is the only place where Monterey pine and Monterey cypress grow together with Bishop pine (*Pinus muricata*).

Remaining native pine and cypress forests are threatened by urban development, genetic contamination from horticultural stock, urban fire management practices, disease, and invasive non-native species. Native stands are protected at Point Lobos State Reserve, Jack's Peak Regional Park, and privately owned Huckleberry Hill S.F.B. Morse Preserve, which protects an inland pygmy forest growing on acidic hardpan soils derived from an ancient marine terrace.

Local citizens have formed a Monterey Pine Forest Watch to foster awareness and appreciation of the Monterey pine forest. A conservation plan, with funding to come from federal, state, and local conservation sources, has been initiated to protect the remaining pine and cypress forest.

Nicole Nedeff and Mary Ann Matthews

Fog-shrouded and windswept in their native habitat, LEFT, *Monterey pine* (**Pinus radiata**) *and Monterey cypress* (**Cupressus macrocarpa**) *are now successfully cultivated throughout the world. The few remaining natural stands along the central coast of California face serious decline.* • *The closed cones of Monterey pine,* TOP, *retain viable seed for many years and require mechanical breakage or the heat of a fire to open. Following fire, seeds are readily shed.* • *The rare Hickman's onion* (**Allium hickmanii**), ABOVE LEFT, *is part of the unique understory that characterizes Monterey pine and Monterey cypress forests.* • *The beauty of Monterey cypress trees at Point Lobos,* RIGHT, *attracts visitors from all over the world.*

BIG SUR COASTAL BLUFF SCRUB

Showy flowers of the coastal bluff scrub community paint the coastal terraces of Monterey's Big Sur coast and form a colorful palette throughout spring and summer months. The coastal terraces of Big Sur were formed as wave-cut platforms. Through tectonic uplift and changes in sea level, the platforms now lie atop steep cliffs about forty feet above the present ocean. Soils are slow to form on these platforms and are shallow and rocky. Winters are cool and wet, summers are long and dry but moderated by fog, and plants are exposed to almost continuous salt-laden winds.

Little Sur manzanita (*Arctostaphylos edmundsii*) and Monterey Indian paintbrush (*Castilleja latifolia*), both rare plants, grow only along the Big Sur coast. Dune buckwheat is the exclusive larval food plant of the endangered Smith's blue butterfly. These rare species live in the rare northern coastal bluff scrub community, found only on undisturbed coastal terraces at scattered locations between Cape Mendocino and Point Conception.

Residential and commercial development, competition from exotic plants, and erosion caused by uncon-

trolled access are threats to this plant community. Fortunately, California has set aside some of its most spectacular coastline in Big Sur as parkland. Garrapata and Andrew Molera state parks protect many of the natural, cultural, and scenic values of the Big Sur coast. But even within the state parks, coastal bluff scrub remains vulnerable to exotics such as iceplant (*Carpobrotus* spp.). Terrace soils are fragile and easily eroded by foot traffic. Park management attempts to route visitors onto designated trails to protect this increasingly rare coastal plant community.

Cynthia L. Roye

Tucked along bluff edges seaward from a low, dense coastal bluff scrub community, LEFT, *bright red flashes of Monterey Indian paintbrush* (Castilleja latifolia) *accent the coastal palette.* • *Monterey Indian paintbrush*, ABOVE, *is uncommon and scattered but locally abundant in good years.* • *The rare Little Sur manzanita* (Arctostaphylos edmundsii), TOP RIGHT, *known from fewer than ten occurrences, is threatened by foot traffic and competition from non-native plants.* • *Springtime on coastal bluffs and mesas*, BELOW, *is a golden time with washes of goldfields* (Lasthenia *sp.*) *and tidy-tips* (Layia platyglossa).

SANTA LUCIA MOUNTAINS

The rugged Santa Lucia Mountains, rising to almost 6,000 feet, extend along the coast from Monterey to San Luis Obispo. Besides providing exquisite views of steep, rocky slopes, cliffs, and waterfalls, this remote and uninhabited range shelters an impressive diversity of plants. The coast redwood (*Sequoia sempervirens*) reaches its southernmost limit here.

Winter storms can dump sixty inches of rain and snow on the west side, sometimes pushing portions of Highway 1 into the sea, while the east side averages only fourteen inches. Summer fog moderates temperatures on the western slopes while eastern slopes can reach 100 degrees F.

A number of rare and unique plants occur within the Santa Lucia Mountains. Bristlecone fir (*Abies bracteata*), also called Santa Lucia fir, is one of the rarest firs in North America. Its lovely weeping forms can be found only in a narrow band roughly fifty-six miles

long and twelve miles wide. Common names bear witness to the restricted ranges of other endemic plants, including Santa Lucia lupine (*Lupinus cervinus*), Santa Lucia gooseberry (*Ribes sericeum*), Santa Lucia bedstraw (*Galium clementis*), and Santa Lucia manzanita (*Arctostaphylos luciana*).

Cattle grazing, commercial and residential development, road maintenance, and recreational activities have taken a heavy toll on the natural resources in some areas of the mountains. Invasive exotic plants are spreading and eliminating native plants.

The Los Padres National Forest has established four botanical areas (Lion Den Spring, Alder Creek, Southern Redwood, and Cuesta Ridge), three wilderness areas (Ventana, Silver Peak, and Santa Lucia), and one research natural area (Cone Peak Gradient) to protect the natural resources. Some of the best wildflower displays grace the U. S. Army's Fort Hunter Liggett Military Reservation at the eastern edge of the Santa Lucia Mountains.

Karen C. Danielsen

Condors once soared above the peaks of the distant Santa Lucia Mountains, OPPOSITE TOP. • *Hairs of stinging lupine* (Lupinus hirsutissimus), TOP LEFT, *can cause blisters.* • *Jolon brodiaea* (Brodiaea jolonensis), TOP RIGHT, *grows on clay soils of grasslands on the east side of the Santa Lucia Mountains.* • *Bristlecone fir* (Abies bracteata), ABOVE, *known from a fossil Miocene flora in western Nevada, is found only in the Santa Lucia Mountains which have served as a refugium.* • *Long a mystery in botanical circles, this population of Muir's raillardella* (Raillardiopsis muirii), BELOW LEFT, *is disjunct with populations in the southern Sierra Nevada, also found on granitic soils.*

99

SAN BENITO MOUNTAIN

Possibly the most unusual feature of San Benito Mountain is the moonscape-like serpentine barrens, which Willis Linn Jepson described as "as barren as one's hand, not even herbaceous vegetation." San Benito Mountain (5,241 feet) near the San Benito-Fresno County line, is the largest and highest elevation block of serpentine in the South Inner Coast Ranges. It is both a serpentine island and a montane outpost. As one of few summits rising above 4,000 feet in the inner south coastal mountains, San Benito Mountain harbors relictual montane plants, isolated since the Pleistocene epoch.

A unique mixed conifer forest of Jeffrey pine *(Pinus jeffreyi)* associated with Coulter pine (*P. coulteri*), gray pine (*P. sabiniana*), and incense cedar (*Calocedrus decurrens*) occurs on the mountain. Jeffrey pine is disjunct over 150 miles from the nearest populations in the Santa Inez Mountains in Santa Barbara County; it forms scattered hybrids with Coulter pine.

San Benito Mountain was origi-

nally designated as a national forest in 1907, but the trees grew too slowly for commercial use. Today this 40,000–acre serpentine area is managed by the Bureau of Land Management. Unfortunately, it is also a popular off-road vehicle use area with hundreds of miles of old mining roads and extensive serpentine barrens. Unrestricted motorcycle use of the barrens and camping on riparian serpentine terraces along Clear Creek have caused much erosion and have eliminated much suitable habitat. A new BLM plan has recently been developed to limit off-road vehicle use. Successful implementation of the management plan would add significant protection to the region's unusual serpentine flora.

Dean Taylor

Bare expanses of serpentine talus, rich in asbestos and cinnabar, and park-like stands of Jeffrey pine (Pinus jeffreyi), LEFT, *distinguish San Benito Mountain, a place unlike any other. Three plants are endemic to the San Benito Mountain serpentine and are found nowhere else: the yellow-fading-to-red San Benito evening primrose* (Camissonia benitensis), *rayless tidy-tips* (Layia discoidea), *and Guirado's goldenrod* (Solidago guiradonis). • *The lovely Brewer's clarkia* (Clarkia breweri), TOP LEFT, *is often found on serpentine soils, and is threatened by cattle grazing.* • *Club-haired mariposa lily* (Calochortus clavatus *var.* clavatus), TOP RIGHT, *is generally found on serpentine soils.* • *Jeffrey pine and gray pine* (Pinus sabiniana) *grow together on serpentine barrens,* ABOVE.

ARROYO DE LA CRUZ

Plant enthusiasts permitted to explore the eleven-square-mile Arroyo de la Cruz, a largely undisturbed area of Hearst Corporation ranchland in northern San Luis Obispo County, count themselves among the lucky.

The Arroyo ranges from flat coastal terraces to gently sloping, rounded hills as high as 600 feet, dissected by steep canyons. A coastal stream bisects the area, which is isolated from similar lowland areas to the north by the steep cliffs along the Big Sur coast. The climate is dominated by marine influences, especially the strong prevailing winds from the northwest.

In this geologically complex area, changes in shoreline profiles during the Pleistocene epoch produced a graduated series of marine terraces that have fine-grained, highly weathered clays and clay-loams that bake hard and crack in summer. Erosion on steep slopes, particularly in canyons, has exposed younger, less weathered soils

Entirely confined within the historic Hearst Corporation's San Simeon Ranch, Arroyo de la Cruz, OPPOSITE TOP, *supports more than 520 species of plants, the richest assemblage in San Luis Obispo County.* • *Five plant species, Arroyo de la Cruz mariposa lily* (Calochortus clavatus *ssp.* recurvifolius), TOP, *dwarf goldenstar* (Bloomeria humilis), MIDDLE LEFT, *Hearst's manzanita* (Arctostaphylos hookeri *ssp.* hearstiorum), LEFT, *maritime ceanothus* (Ceanothus maritimus), *and Hearst's ceanothus* (C. hearstiorum), *grow nowhere else in the world.*

and bedrock. All of this has resulted in a great diversity of plants.

Arroyo de la Cruz is managed by the Hearst Corporation, which maintains grazing at low levels, allowing many native plants to flourish. Perennial grasses, bulb-forming plants, and other native species thrive on marine terraces where plowing has not occurred. Because the Hearst Corporation limits access to the property, much of the destruction that has occurred in nearby areas as a result of off-road vehicle use has been prevented. The conversion of natural chaparral and coastal scrub areas to rangeland and other agricultural uses remains a threat to rare and endangered plants.

David J. Keil

COAST LIVE OAK

Most coast live oak (*Quercus agrifolia*) stands are found within a sixty-mile coastal swath from the Sonoma-Mendocino County border to upper Baja California, Mexico. Unlike many broad-leaved evergreen species, coast live oak is not greatly drought tolerant. Its evergreen habit is believed to have developed in response to mild winters along the coast where summer and winter fog moderates temperatures and reduces moisture loss. Leaves are rarely damaged by frost, and photosynthesis can continue during the winter when soil moisture is highest.

The mature coast live oak is a stout tree growing from thirty to ninety feet tall with a broad canopy, sometimes spreading as much as 150 feet across. In windy coastal locations growth is contorted with low, undulating branches reaching across the ground. The tree is distinguished by dark green holly-like leaves and smooth grayish bark that becomes ridged and covered with lichens in older specimens. Out of hundreds of animal and thousands of invertebrate species that depend on the coast live oak for food or shelter, three are particularly significant: the oak moth, the acorn woodpecker, and the arboreal salamander. The oak moth caterpillar grazes on the leaves, sometimes completely defoliating trees but rarely killing them; the woodpecker consumes quantities of caterpillers and collects and stores acorns in spectacular grana-ries; and the arboreal salamander, not obvious to the casual observer, lives high in tree cavities and forages at night for insects and invertebrates.

Pamela C. Muick

The dark green foliage of coast live oak (Quercus agrifolia) *against the summer gold of plains, valleys, and foothills characterizes California's coastal landscape. In the past, twelve Native American tribes depended upon acorns of coast live oak as a dietary staple. Today, approximately two-thirds of all Californians, almost twenty million people, live within the distribution of this handsome live oak. Some one hundred ordinances have been written by municipalities in an effort to preserve the oaks.*

MORRO BAY

Morro Bay was formed when rising waters of the Pacific submerged the valleys of Chorro and Los Osos creeks and formed a barrier beach, closing off the tidal lagoon from the ocean. The extraordinary views of Morro Bay and Morro Rock have spawned the communities of Morro Bay, Baywood Park, and Los Osos. Yet much of the bay's shoreline remains undeveloped as part of Morro Bay State Park. Active dunes along the sand spit support the threatened beach spectacle-pod (*Dithyrea maritima*) and three sand-verbenas, *Abronia maritima*, *A. latifolia*, and *A. umbellata*, in a progression from foredunes to backdunes.

Stabilized sand dunes, which abut the southern end of Morro Bay, support a large population of Morro Bay manzanita (*Arctostaphylos morroensis*), a tree-sized shrub known from only about 840 acres here and nowhere else. Morro Bay's coastal dune scrub, itself a rare community, is home to several rare plants and the more common beach silver lupine (*Lupinus chamissonis*), which hosts the rare Morro Bay blue butterfly. The dunes also provide habitat for the most endangered mammal in North America, the Morro Bay kangaroo rat.

A plan to establish a permanent urban boundary or "greenbelt" around the communities of Baywood Park and Los Osos is being combined with a conservation planning program in the area. The conservation plan's goal is to permanently protect the plants, animals, and habitats unique to this scenic area while allowing for further development. Partners in this effort include local citizens, conservation groups, agencies, and business interests.

David Chipping

Morro Bay's tidal mudflats, RIGHT, *provide important habitat for migrating birds, and the bay's watershed, one of the biologically richest in California, is home to more than two dozen rare plants, animals, and natural communities. • A large population of rare salt marsh bird's-beak* (Cordylanthus maritimus *ssp.* maritimus), TOP LEFT, *occurs at the Sweet Springs Nature Preserve. • Morro Bay is the last known habitat in the world for sea-blite* (Suaeda californica), TOP RIGHT, *which is found along the lagoon's high-tide line.*

CUESTA RIDGE WEST

A unique Sargent cypress (*Cupressus sargentii*) forest dominates Cuesta Ridge West, the serpentine barrens that form the crest of the Santa Lucia Range north of San Luis Obispo. The forest extends from Cuesta Peak northwest along the ridge, a distance of about three miles.

Cuesta Ridge West extends northwest from Highway 101 at Cuesta Pass to Cerro Alto Peak near State Highway 41. It ranges between 2,000 and 2,600 feet in elevation, and is composed mostly of serpentine or serpentine-derived soils. The climate is strongly influenced by the nearby Pacific Ocean.

Recently, U.S. Forest Service botanists installed barriers to protect the rare plant populations from off-road vehicles. Every semester, local colleges use Cuesta Ridge West as an outdoor classroom for students studying serpentine endemics, rare plants, and fire ecology. They are at present studying the west ridge following a major wildfire in August 1994 that burned a mosaic of vegetation. Students will have data to compare pre-fire and post-fire occurrences of numerous plant species.

Malcolm McLeod

The Sargent cypress forest (Cupressus sargentii), ABOVE, on Cuesta Ridge has been designated a Botanical Area by the Los Padres National Forest, providing the cypress and several rare plants with special protections. • Cones of Sargent cypress, TOP, like those of other cypresses, are tightly sealed until great age or fire causes them to open and release their seeds. • Cuesta Pass checkerbloom (Sidalcea hickmanii ssp. anomala), LEFT, is one of several rare plants found along the windswept ridge and through the cypress forest. • The rare Pecho manzanita (Arctostaphylos pechoensis), BELOW, is restricted to siliceous shale in coastal mountains of San Luis Obispo.

THE NIPOMO-GUADALUPE DUNES

The Nipomo-Guadalupe Dunes constitute one of the largest dune systems in California, occupying a nearly continuous lowland strip from Pismo Beach in southwestern San Luis Obispo County to Mussel Rock in northwestern Santa Barbara County. The dunes extend nearly five miles inland onto old elevated marine terraces. The Santa Maria River drains the southern portion of the dunes, and several smaller streams, lakes, and ponds form localized wetlands within the dune complex.

Plant life on the Nipomo-Guadalupe Dunes is diverse, with more than 400 species, subspecies, and varieties occurring in coastal strand, dune scrub, marine dune chaparral, coastal live oak woodland, riparian woodland, freshwater marsh, and dune swale communities. More than a dozen plants of the dunes are considered rare, threatened, or endangered.

The Nipomo-Guadalupe Dunes occupy a patchwork of privately and publicly owned land. Public lands include Pismo State Beach, Pismo State Vehicular Recreation Area, and Rancho Guadalupe Dunes County Park. Off-

The Nipomo Dunes, BELOW, include marine terraces and sand deposits of various ages, the result of changes in sea level and shoreline profile during the Pleistocene and Holocene epochs. Except where disturbed by human activities, these ancient dunes are largely vegetated. • Two threatened plants of the Nipomo-Guadalupe Dunes are La Graciosa thistle (Cirsium loncholepis), LEFT, occurring in dune swales and wetlands, and surf thistle (C. rhothophilum), ABOVE, scattered in the leeward side of the foredunes here and in the dunes of Vandenberg Air Force Base to the south. • Giant coreopsis (Coreopsis gigantea), TOP LEFT, creates yellow bouquets among the spectacular sand dunes. • The rare crisp monardella (Monardella crispa), TOP RIGHT, spreads by rhizomatous roots but is threatened by off-road vehicle activity.

road vehicle use in some of these areas has removed most of the plant cover from the dunes, creating active dunes that are now advancing across vegetated areas and burying everything in their path. The Nature Conservancy has recently undertaken management of the Nipomo Dunes Preserve, which includes both privately and publicly owned portions of the dunes. The Nature Conservancy has eliminated recreational vehicle use on its preserve.

David J. Keil

107

GREAT VALLEY

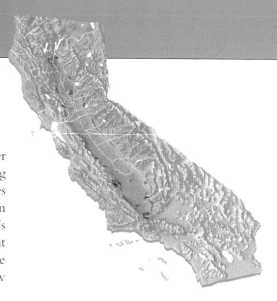

California's Great Valley, bordered on the east by the Sierra Nevada and on the west by the Coast Ranges, was born in the sea. Its ancient soils, filled with fossilized remains of fish, whales, and dolphins that flourished more than 100 million years ago, accumulated while the valley lay under shallow seas. Today, a 400–mile–long plain sprawls over 20,000 square miles where sediment has accumulated in places to a depth of ten miles. Today's Great Valley plain, with its nearly flat topography, is broken only by the volcanic Sutter Buttes and the shallow

grassland. The Nature Conservancy has acquired and protected vernal pools and valley grasslands at the Jepson Prairie near Fairfield and Vina Plains north of Chico. The Conservancy has purchased examples of riparian forests at the Cosumnes and Kaweah rivers and is working with the California Department of Fish and Game and the Bureau of Land Management to jointly manage a major preserve, the Carrizo Plain Natural Area. In Merced County, thousands of acres of grasslands and wetlands are protected within a grasslands ecological area. These fragments of the valley's once-rich botanic heritage can, with appropriate management, provide a glimpse of California's past for future generations.

Joseph L. Medeiros

engravings of two major river systems, the Sacramento and the San Joaquin. The Tulare Lake Basin lies landlocked at the valley's southern end.

The Sierra Nevada regularly receives heavy snowfalls that, in the past, melted and flowed across the valley as wild and cold rivers. Away from the reaches of floodwaters, the land was watered only by rainfall. Here, millions of acres of prairie formed. Perennial and annual grasses and myriad wildflowers sprang forth like an artist's palette of color and texture. Vernal pools dotted the landscape. The vast mosaic of forests, marshes, and grasslands teemed with wildlife—elk, deer, pronghorn antelope, grizzly bears, mountain lions, and wolves; winter skies were darkened by ducks, ibis, cranes, geese, swans, and pelicans.

Remnants of this once-abundant valley habitat have been protected in a few preserves, though these are significantly altered by weedy species that have arrived from other continents during the last few centuries. Caswell State Park protects a small fragment of riparian forest; the Great Valley Grasslands State Park shelters a remnant

In the Great Valley, water runoff slows, meanders, and floods, sculpting a lacework of life-sustaining wetlands, prairies, and woods, LEFT. *Millions of acres of freshwater marsh once flourished, sheltering countless species of plants and animals. Oaks, cottonwoods, alders, and willows crowded the floodplains in towering riparian forests. Today, little of the valley's once-rich natural ecosystem remains, and crops and cities dominate the land. • Lindley's blazing star* (Mentzelia lindleyi), TOP LEFT, *joins the California poppy as one of the state's few truly orange-flowered native plants. • Long thought to be extinct, one population of veiny monardella* (Monardella douglasii ssp. venosa), TOP RIGHT, *was rediscovered in 1992. It remains severely threatened by development. • Lupines* (Lupinus spp.) *and fiddlenecks* (Amsinckia spp.), ABOVE RIGHT, *once carpeted the valley floor in spring.*

NORTH TABLE MOUNTAIN

orth Table Mountain lies in the Sierran foothills north of Oroville, in the center of Butte County. This flat but highly dissected mesa is an exposed portion of the early Oligocene basaltic Lovejoy Formation, a lava flow that at twenty-three million years is California's oldest. The black rock is fractured and eroded, giving rise to numerous intermittent streams—many produc-

ing spectacular ribbon waterfalls at edges of the table in winter and spring. Typical foothill trees, such as gray pine (*Pinus sabiniana*) and interior live oak (*Quercus wislizenii*), encircle the mesa below the fragmenting vertical cliffs of basalt and extend up into the eroded ravines, but the top of the mesa is nearly treeless. Table Mountain is covered with grassland, predominantly of annuals, which is quite similar to that

lying 1,000 feet lower in the Sacramento Valley to the west.

Flower patterns on the tabletop reflect an array of habitats, ranging from xeric, unfractured bare rock to deep, moisture-retaining soils and even vernal pools. The flat, dry expanses of basalt harbor pink early-flowering and succulent bitterroot (*Lewisia rediviva*) and purple wild garlic (*Allium cratericola*). Edges of craggy basalt outcrops

host yellowish jewelflowers (*Streptanthus diversifolius* and *S. tortuosus* var. *suffrutescens*—the latter typical of much higher elevations).

This grassland is less disturbed by human activities than its Great Valley counterpart, and contains fewer introduced species. Despite a long history of moderate cattle grazing here, the mountain's vegetation represents an important remnant of pre-settlement grassland. In 1993 the California Department of Fish and Game purchased 3,372 acres of this well preserved grassland, an area that includes the rare northern basalt flow vernal pools. The overall scenic values of the

mountain, and the botanical and geological features that make it unique, are included within the North Table Mountain Wildlife Area.

Robert A. Schlising

Table Mountain, FAR LEFT, *is known for its spring wildflowers, particularly sheets of California goldfields (Lasthenia californica). • Phantom Falls,* TOP RIGHT, *cascades over basalt blocks. • Late in the season, flowers of Douglas's violet (Viola douglasii),* ABOVE LEFT, *remain unopened and self-pollinate to ensure seed set. • Kellogg's monkeyflower (Mimulus kelloggii),* MIDDLE LEFT, *commemorates Albert Kellogg, one of the founders of the California Academy of Sciences. • One of the delights of spring is finding Lobb's poppy (Eschscholzia lobbii),* ABOVE RIGHT, *in the tabletop garden. • Wild garlic (Allium cratericola),* LEFT, *lodges in Table Mountain's rocky nooks and crannies.*

VERNAL POOLS OF THE GREAT VALLEY

Much of the grassland and blue oak woodland of the Great Valley was once dotted with millions of vernal pools. These seasonal wetlands develop in shallow depressions where the underlying hardpan or claypan acts as a barrier to downwardly percolating water. Rain collects in winter and, as water evaporates through the spring, massed populations of specialized plants produce unsurpassed displays of floral color.

Introduced Mediterranean annual plants, which have so pervasively replaced the native flora in grasslands throughout the state, have not gained a foothold in the valley's vernal pools, which still show us glimpses of pre-agricultural California. But the vernal pools are more than colorful displays of native wildflowers; they are small eco-systems that support a complex web of plants and animals ranging from microscopic algae to myriad invertebrates, frogs, salamanders, and waterfowl.

Between sixty-six and ninety percent of this Great Valley habitat was lost to agricultural development, mineral extraction, and urban expansion by the 1970s. Coupled with the high level of speciation found in vernal pools, this has resulted in a great many vernal pool plants and animals becoming rare, threatened, or endangered. Several vernal pool species are now known from fewer than ten populations, and others are restricted to a single county.

Nevertheless, some natural vernal pools can still be found in every county in the Great Valley. Many vernal pools persist on private cattle ranches where sustainable grazing practices are compatible with the maintenance of these interesting ecosystems. The Nature Conservancy has established large vernal pool preserves at Jepson Prairie in Solano County and the Vina Plains in Tehama County. The Department of Fish and Game protects vernal pools at its Table Mountain Wildlife Area in Butte County, Big Table Mountain in Fresno County, and several smaller reserves.

Robert Holland

VERNAL POOL SPECIALISTS: BEES

Easily overlooked in the flowery vernal pools of springtime are tiny native bees and other small insects that visit these flowers for pollen and nectar. One such pollinator is the ground-nesting native bee, *Andrena (Diandrena) blennospermatis*, a specialist dependent on flowers of yellow carpet (*Blennosperma nanum*). Females visit only flowers of yellow carpet for pollen, and fly only when their host flowers are in bloom. Following mating with a male bee, each female digs a shallow hole in an adjacent upland area and constructs a brood chamber in the soil. She then collects pollen from yellow carpet and returns to the brood cell, where she shapes the pollen into a small ball of dough-like consistency by adding some nectar. She lays an egg on top of the pollen provisions and seals the cell. The pollen ball provides each offspring with all the food it needs to complete its development, and the mother bee has no further contact with her young. In autumn, the offspring transform to adults, but remain in the brood cell until spring, when they emerge as pollen host plants begin to bloom. Exactly how the life cycles of the bees and plants are so closely synchronized is not well understood, but the plant is pollinated and sets seed, and the bees nourish themselves and their young.

The close synchrony of the annual life cycles of these pollen-specific bees and their host plants suggests that this interdependence is essential to the well-being of the vernal pool community. The bees live not in the pools but in nearby uplands, and have limited foraging ranges. Therefore, preservation efforts and mitigation plans that try to compensate for losses to vernal pool habitats should consider the ecological interactions between pools and upland areas.

Robbin W. Thorp

The tiny solitary bee, Andrena (Diandrena) blennospermatis, TOP, *visits a flower of its host plant, yellow carpet* (Blennosperma nanum), ABOVE, *that relies upon specific bees for pollination.*

Because of the isolation of individual vernal pools and the small range of bee pollinators, several genera of plants, such as meadowfoams (Limnanthes) *and* Downingia, *have evolved into a dozen or more species. Douglas's meadowfoam* (Limnanthes douglasii) *creates white rings around drying pools,* TOP LEFT. • *Concentric rings of white, yellow, pink, and sky-blue flowers successively line the drying rims of the pools at Phoenix Field,* FAR LEFT. *By mid-summer pool bottoms are baked hard, and the prolific abundance of annual flowers and invertebrates has all but disappeared; they will reappear with next year's winter rains.* • *Each spring around mid-April, when 2,000 acres of wildflower grasslands and vernal pools on the upper terrace soils of the Sacramento Valley reveal their magic, the minty aroma signals the presence of Douglas's mint* (Pogogyne douglasii), LEFT. • *Seedlings of the distinctive rare Colusa grass* (Neostapfia colusana), ABOVE LEFT, *shown here in full flower, typically flourish only in large, deep pools.*

SUTTER BUTTES

The 400-mile-long trough of the Great Valley is punctuated by only one source of high relief, the Sutter Buttes. Strikingly symmetrical, this circular cluster of Pleistocene volcanoes has been compared to a medieval castle. It has a rampart in the form of a peripheral ring of volcanic debris; a moat provided by an inner valley of upturned sedimentary beds of sandstones, shales, and gravels; and a central castle represented by the craggy interior of andesite and rhyolite domes.

The shallow, rocky soils of the rampart support open grassland on the south and west portions of the ring, and park-like blue oak (*Quercus douglasii*) woodland on the north and east. In the rugged interior, blue oak stands are interwoven with dense chaparral particularly on south-facing slopes, woodlands on north-facing slopes, and patches of grassland. Rocky cliffs, steep slopes, riparian courses, sloping meadows, and vernal pools add to the highly concentrated horizontal and vertical complexity.

On a clear day, the Sutter Buttes, ABOVE, *a miniature mountain range lying between the Sierra Nevada and the Coast Ranges to the west, can be seen from many locations in the Sacramento Valley.* • *Bush lupine* (Lupinus albifrons), TOP, *found in openings in the chaparral, has variable color forms.* • *Mule ears* (Wyethia helenioides), RIGHT, *grows in open sunny places. Their leaves are sometimes densely hairy, a feature that reduces sun exposure.* • *Unicorn plant* (Proboscidea louisianica), OPPOSITE TOP, *is an ill-smelling and uncommon plant in the Sacramento Valley.* • *Stands of blue oak* (Quercus douglasii), FAR RIGHT, *create patterns of breathtaking beauty in valley foothills.*

114

The unique combination of plants and animals here includes disjunct populations growing far from their typical range, such as rock gooseberry (*Ribes quercetorum*); species representing extreme range limits, such as Fremont globemallow (*Malacothamnus fremontii*) and thyme-like pogogyne (*Pogogyne serphylloides*); and others that occur also in the foothills of the Sierra Nevada or in the Inner Coast Ranges, but not both, such as narrow-leaf goldenbush (*Ericameria linearifolia*).

Today, this area is primarily private rangeland, and has not experienced large-scale development. Invasion of exotic species and vulnerability to future development are the primary threats to native plants and wildlife here. Since 1976, field trips to this beautiful area have been offered, first by the Sutter Buttes Naturalists and later by the Middle Mountain Foundation. The Middle Mountain Foundation has been working with landowners in the Buttes and public agencies to develop easements and zoning that will protect long-term agricultural and natural uses.

Walt Anderson

THE SACRAMENTO-SAN JOAQUIN RIVER

At the confluence of the two great rivers draining the western Sierra Nevada and Inner Coast Ranges is a vast and complex mosaic of water, marsh, and islands. These mighty rivers, the Sacramento flowing from the north and the San Joaquin from the south, together with their tributaries, continuously alter the landscape as they slow down and deposit their mineral and organic particles before flowing into San Francisco Bay. The accumulation of these deposits over thousands of years has created an extensive delta. Nearer to San Francisco Bay, tidal fluctuations cause diurnal and seasonal incursions of salt water and create a wetland gradient from freshwater to brackish to saltwater marshes.

Historically, the delta wetland consisted largely of freshwater marsh vegetation, with oak, willow, and cottonwood riparian forests. Because of the nutrient-rich organic soils and ready availability of water, many of the delta islands have been converted to agriculture over the past 100 years. Around the turn of the century, levees were built up to protect these islands, but, over the years, the highly organic soil disintegrated as it dried. The centers of the islands are gradually subsiding below sea level and today require extensive draining and pumping to be retained for agricultural use.

Natural vegetation can now be found only in discontinuous patches, generally on the outer levee banks or on berm islands. Few oaks or cottonwoods remain. In an attempt to preserve the islands, native trees have been

DELTA

replaced around the perimeters by miles of riprap.

Of the approximately 11,000 square miles of delta, the only formal plant protection is provided at Calhoun Cut, where about 1,000 acres have been designated as a California Department of Fish and Game (DFG) Ecological Reserve. DFG's Grizzly Island Wildlife Area protects a large amount of natural marsh habitat, though habitat is not protected specifically for rare and endangered plants.

Niall F. McCarten

The confluence of two great rivers, the Sacramento and the San Joaquin, create an immense delta, OPPOSITE, *where sediments rich in nutrients are deposited. The deepest alluvial soils support great riparian forests. • The rare rose-mallow (Hibiscus lasiocarpus),* TOP, *is threatened by ongoing alterations of the delta for agriculture and development. • Stands of cattails (Typha latifolia),* ABOVE, *occur in freshwater marshes, where they attract great flocks of red-winged blackbirds. • Mason's lilaeopsis (Lilaeopsis masonii),* LEFT, *a grass-like member of the carrot family, carpets locally exposed eroding mud banks adjacent to freshwater marshes. It is one of eight plant taxa in the delta considered rare today.*

117

VALLEY OAK FOREST

Valley oak (*Quercus lobata*), the monarch of deciduous oaks of the West, grows only in California, where it is the dominant tree in a structurally diverse forest. It grows only on deep soils, often where there is a high water table and, typically, on the rich alluvial soils of the floodplains close to the rivers of the Great Valley and Coast Ranges. The valley oak forest is distinguished by its open grandeur. The trees often stand well apart, as if they had been planted in a park, because their roots compete for water.

Because of the tremendous agricultural potential of the Great Valley's deep alluvial soils, much of the valley oak riparian forest was cleared by early settlers soon after the Gold Rush. Subsequent water diversions, groundwater pumping, building of levees to

Valley oak forest, LEFT, once dominated more than a million acres of the Great Valley, and supported more species of wildlife than any other type of forest in California. Today, only degraded remnants of those ancient, mighty forests remain. • During the last half of its naturally long life of 300 years or more, the valley oak assumes a characteristic weeping silhouette, densely leafy and often festooned with lichens, galls, and wild grape, TOP. • The deeply lobed leaves of valley oak emerge in late spring in shades of pink and bronze, remain green through the summer, and then turn russet brown and fall, leaving characteristic silhouettes as winter rains arrive, ABOVE.

restrain floodwaters, and damming of most of the major rivers caused further decline of the species and its habitat.

Most landowners today, including ranchers, welcome the valley oak on their land and are willing to modify their management practices to ensure its health and vigor. Yet the lack of young saplings suggests that many of the remaining stands of valley oak are not regenerating. If the valley oak forest is to survive, research must focus on the conditions under which seedlings take root and saplings survive.

All remaining valley oak forests are significantly degraded. The best examples of these forests, once so magnificent and extensive, can be seen near Chico in Bidwell River Park along the Sacramento River and also at Bidwell Park along Chico Creek, at Woodson Bridge State Recreation Area in Corning, at Brushy Lake behind Cal Expo in Sacramento, in Caswell State Park south of Manteca, at The Nature Conservancy's Cosumnes River Preserve near Sacramento, and at the Conservancy's Kaweah Oaks Preserve near Visalia.

Tom Griggs

RESTORING VALLEY OAKS

At its Cosumnes River Preserve in southern Sacramento County, The Nature Conservancy of California is working with dedicated volunteers to restore a valley oak riparian forest along flood-prone lands that had been previously cleared. The Cosumnes River is one of the few tributaries of the Sacramento without major upstream dams, so it still maintains a near-natural flooding regime. In fact, periodic flooding is the main ecological process that rejuvenates the riparian forest.

The first planting of valley oak acorns and Oregon ash (*Fraxinus oregona*) seed was carried out at the Cosumnes River Preserve on twelve acres at the visitor trailhead in 1988. As the oaks mature and create an overhead canopy, seeds of other riparian shrubs and trees, dispersed by floods or wildlife, are expected to become established, and the natural diversity typical of a riparian forest should be restored through a natural successional process.

The survival and growth of seedlings are being monitored as they respond to soil differences, weed competition, and herbivore browsing. Factors such as the increased rodent populations that accompanied the spread of non-native annual grasses, as well as lowered water tables from extensive groundwater pumping, are being examined in relation to seedling survival. The program will emphasize restoration of the diversity found in a natural valley riparian community.

The commitment of hundreds of citizens from surrounding communities helps bring the return of a valley oak forest closer to reality. Valley oak forests were once one of the most majestic sights in the Great Valley of California.

Tom Griggs

Valley oaks (Quercus lobata), *growing on the floodplain of the Cosumnes River, attain their great size by tapping into stored groundwater.*

GRASSLANDS OF THE SAN JOAQUIN VALLEY

Expansive seas of grasses in the San Joaquin Valley once were home to thousands of species of native plants and animals, linked in elaborate food webs that pulsed with life. Deer and pronghorn grazed valley grasslands only a few hundred years ago, and carnivores such as grizzlies, wolves, coyotes, foxes, and raptors were common. The grizzlies and wolves are gone. Some calculate that less than five percent of the original native grassland acreage remains. Today, exotic Mediterranean annual grasses dominate these lands, while forty-seven endangered, threatened, or sensitive plant and animal species are hanging onto existence here.

San Joaquin Valley grasslands rely upon sparse winter rains for water. The scorching summer heat and drought forge evolutionary strategies for survival. Hardy perennial bunchgrasses grow during moist times and survive summer by storing food in deep roots, while annual grasses and wildflowers hastily flower during balmy spring days and set seed before they die. They oversummer as seed.

Early settlers quickly recognized the value of the grasslands and their rich, deep soils. Later, the valley was plowed and dry-farmed for grains. Dams and irrigation augmented scarce surface water. The grasslands were converted to grow food and fiber and before long, most of the wild prairie was gone.

Ambitious efforts to reassemble pieces of the past continue. An initially modest attempt in 1972 to preserve San Luis Island in Merced County, an island of grasslands and wetlands between the San Joaquin River and Salt Slough, has resulted in a major interagency and private effort that today manages and protects a 160,000-acre grasslands ecological area. There are plans for restoring this great expanse of grassland to a more biologically diverse and natural condition.

Joseph L. Medeiros

Some years, spring brings a brilliant carpet of lupines and purple owl's-clover (Castilleja exserta *ssp.* exserta), ABOVE RIGHT, *to the valley floor.* • *The San Joaquin Valley grasslands,* RIGHT, *could once be compared to the Great Plains of North America or even the Serengeti of East Africa. Today, vestiges of San Joaquin Valley grasslands remain primarily in floodplains or in highly alkaline soils that cannot be effectively farmed. On these remnants are found endangered plant communities such as alkalai sacaton grasslands dominated by the coarse native perennial grass alkali sacaton* (Sporobolus airoides), *shown here, as well as valley saltbush scrub, iodine bush scrub, and vernal pools.* • *Aspen onion* (Allium biseptrum), LEFT, *occurs in Troy Meadows on the Kern Plateau.* • *The bright orange fiddlenecks* (Amsinckia *spp.*), MIDDLE RIGHT, *can be poisonous to cattle.* • *With more than seventy species and dozens of varieties, lupines* (Lupinus *spp.*), FAR RIGHT, *are difficult to identify.*

CARRIZO PLAIN

Tucked away in a valley in the Inner Coast Range between Bakersfield and the coast in southeastern San Luis Obispo County, the Carrizo Plain stretches from the Elkhorn Hills west to the Caliente Range.

The valley floor reaches its lowest point in Soda Lake, an alkaline sink that dries during most summers. It is ringed by three natural communities, each with a dominant plant species: valley sink scrub with iodine bush (*Allenrolfea occidentalis*), valley allscale saltbush scrub with allscale (*Atriplex polycarpa*), and spiny saltbush scrub with spiny saltbush (*Atriplex spinifera*). A localized population of the pale yellow peppergrass (*Lepidium jaredii* ssp. *jaredii*) blankets an alkali flat south of Soda Lake; this rare species is known only from one other population, near Devil's Den in Kern County. Two other rare San Joaquin Valley endemics are found near Soda Lake—lavender alkali larkspur (*Delphinium recurvatum*) and Lost Hills saltbush (*Atriplex vallicola*).

Upslope, the influence of the alkaline soils is largely left behind. Grasslands dominated by pine bluegrass (*Poa scabrella*) and various herbaceous species blend into shrublands of widely spaced Mormon tea (*Ephedra californica*), goldenbush (*Ericameria hooveri*), and California juniper (*Juniperus californica*). Here are scattered populations of the endangered California jewelflower (*Caulanthus californicus*), known only from the Carrizo Plain, the hills of southwestern Fresno County, and the Cuyama Valley of northern Santa Barbara County.

Vegetation on the Carrizo Plain represents a microcosm of that once found in the Great Valley to the east. Some of the native fauna still persist here, including the rare San Joaquin kit fox and San Joaquin antelope squirrel, the blunt-nose leopard lizard, and the giant kangaroo rat.

A 180,000-acre Carrizo Plain Natural Area is managed cooperatively by the California Department of Fish and Game, The Nature Conservancy, and the U.S. Bureau of Land Management. Agencies are working with ranchers to develop appropriate levels of grazing to maintain the floral diversity. The Carrizo Plain Natural Area provides one of the best opportunities

to see a unique complement of rare and endangered plants and natural communities typical of the southern San Joaquin Valley, a vanishing treasure.

Deborah Hillyard

---

*The Carrizo Plain offers a glimpse of California's past; impressive wildflower displays carpet the valley floor in years of good rainfall. Purple owl's-clover (*Castilleja exserta ssp. exserta*), LEFT, is not a clover at all, but a snapdragon, and it often occurs with goldfields (*Lasthenia californica*), a plant that helps define the Golden State. • The rare San Joaquin woolly-threads (*Lembertia congdonii*), TOP, found on sandy flats on the Carrizo Plain, ranges throughout the western San Joaquin Valley and low foothills. • Tidy-tips (*Layia platyglossa*), ABOVE, generally flowers after goldfields, but often in the same area. • Thistle sage (*Salvia carduacea*), BELOW LEFT, is a spiny, annual mint. • In some years, the Temblor Range hills, BELOW, bordering the Carrizo Plains, glow with monolopia (*Monolopia major*), one of about 2,000 California endemic plants.*

SALTBUSH SCRUB OF THE SOUTHERN SAN JOAQUIN VALLEY

Prior to the settlement of California by Europeans, the southern part of the Great Valley was partially covered with shrubby plant communities. As much as 800,000 acres of valley saltbush scrub may have existed in the southern San Joaquin Valley between Fresno and Bakersfield and in the Carrizo Plain. Today, after a century of irrigated agriculture and urban development, fewer than 36,000 acres, just four and a half percent, remain.

Why does saltbush scrub grow in the valley? The answer lies in the valley's desert-like climate and complex soils. The Coast Ranges create a rain shadow over the southern valley, which receives an average of fewer than seven inches of rain a year and experiences frequent droughts. A dense ground fog called "tule fog" develops in the winter, increasing soil moisture, which in turn influences vegetation. Tule fog is the principal climatic difference between the San Joaquin Valley and the Mojave Desert to the southeast.

Fine-textured alluvium carried down from the Coast Ranges contains large quantities of soluble salts and has formed extensive areas of alkali sinks and playas in the valley. To the east, the alkaline soils are overlain by more sandy, non-alkaline alluvium from the Sierra Nevada. The result is a rich mosaic of soils that vary in texture and alkalinity.

There are two major plant associations within the valley saltbush scrub community: valley spiny saltbush scrub is dominated by its namesake, spiny saltbush (*Atriplex spinifera*), and grows on semi-alkaline soils upslope from the wettest lands; valley allscale saltbush scrub is dominated by allscale (*Atriplex polycarpa*) and occurs on non-alkaline soils on the valley floor.

This rare natural community is protected at only a few sites in the southern San Joaquin Valley and on the Carrizo Plain in San Luis Obispo County. Examples of valley spiny saltbush scrub can be seen on the Kern National Wildlife Refuge and the Semitropic Ridge Preserve in Kern County and on the Carrizo Plain Natural Area west of the San Joaquin Valley. Protected allscale saltbush scrub communities occur on Semitropic Ridge and at

Historical accounts of the valley speak of colorful spring wildflowers, but few refer to the valley saltbush scrub community, BELOW, *where low to medium-sized shrubs grow in widely spaced patterns. In years of abundant rainfall, grasses and wildflowers form brilliant displays beneath and between the widely spaced shrubs. Even then, wide expanses of the valley appear monotonous and dry most of the year.* • Goldfields (Lasthenia spp.) *and tidy-tips* (Layia spp.), RIGHT, *grow on the Grassland Ecological Reserve in Merced.* • *The rare California jewelflower* (Caulanthus californicus), ABOVE RIGHT, *has evolved in this arid habitat and grows nowhere else. Only twenty populations remain today.*

Lokern Preserve, also in Kern County. These sites contain most of the high-quality valley saltbush scrub habitat remaining in California.

Roxanne L. Bittman

BAKERSFIELD CACTUS

California's rarest cactus, the Bakersfield cactus (*Opuntia silaris* var. *treleasei*) grows on sandy soils, often with large cobbles and boulders. Temperatures here reach 100 to 110 degrees F. for four months of the year, and rainfall averages only four inches. Good drainage is critical in preventing fungal diseases caused by excess moisture. In late March to May, the bright magenta flowers produce abundant amounts of nectar and pollen, which attract many insects. Because of the extreme fragmentation of its habitat, some botanists are concerned that the remaining populations of Bakersfield cactus may become too small and isolated to sustain populations of insects that are critical for pollination.

Since 1990, when the Bakersfield cactus was listed by the state and federal governments as an endangered species, two of the remaining seventeen populations have been destroyed by development, and another nine are in the path of urban, energy, and agricultural development. Bakersfield cactus is imperiled because its limited range almost coincides with the boundaries of the city of Bakersfield. Bakersfield cactus is protected at Sand Ridge, a several-hundred-acre site owned by The Nature Conservancy that represents only one percent of its original distribution, and also at a 6,000-acre site southwest of Bakersfield, owned by the California Department of Fish and Game.

Roy van de Hoek

In April 1844, Charles Preuss, Fremont's mapmaker, wrote in his journal of the low hills between the Kern River and Caliente Canyon: "The hilly country is completely bleak, without any vegetation except a beautiful species of cactus whose magnificent red blossoms grace this sad sandy desert in a strange manner."

SIERRA NEVADA

The Sierra Nevada reigns over the California landscape. Home to more than half of California's plant species, one-third of which are endemic to these mountains, the Sierra Nevada trends northwest to southeast for nearly 400 miles, narrowing north to south from seventy-five to fifty miles wide.

The highest peak in the lower forty-eight states occurs here, the majestic 14,496-foot tall Mount Whitney. While the range's western slope is a gradual incline, most of the eastern escarpment drops off steeply. The highest peaks are concentrated along the central to southern crest. Although an uplifted mass of intruded granite dominates the range, other geologic formations are important in accounting for the great species diversity and endemism found here. The northern Sierra is predominantly volcanic in origin, connecting to the southern Cascades at Lassen Volcanic National Park. The central and southern Sierra are both mainly granitic, but also have several areas of metamorphic and metasedimentary rock. Yosemite National Park, located in the central Sierra, and Sequoia-Kings Canyon National Parks

in the southern Sierra provide dramatic expanses of granitic rocks.

At the highest reaches of the mountains, subalpine and alpine forests differ in dominant species based on location and distribution. Whitebark pine (*Pinus albicaulis*) in the northern and central Sierra is restricted to an elevational band generally between 8,000 and 12,000 feet and is replaced by foxtail pine (*P. balfouriana*) in the southern Sierra. Douglas-fir is a common associate in the mid-elevation mixed conifer forests of the northern and central Sierra, but is absent from the southern Sierra. The extreme southeastern Sierra Nevada is desert-like, its flora greatly influenced by the adjacent Great Basin, Mojave Desert, desert mountain ranges, and Transverse Ranges of southern California.

Famous examples of the Sierra's endemic species include giant sequoia (*Sequoiadendron giganteum*), growing

Naturalist John Muir described the Sierra Nevada, RIGHT, as "the Range of Light, the most divinely beautiful of all the mountain chains I have ever seen." This spectacular view of the Sierra Nevada Range is from across an alkali lake near Lake Crowley. • Flowers of sticky monkeyflower (Mimulus aurantiacus), TOP, are usually a soft orange color but can vary from white to red. • Quaking aspen (Populus tremuloides), ABOVE, is so called because a narrowing of the leaf petiole causes the leaves to tremble in the slightest breeze. • Brewer's lupine (Lupinus breweri), OPPOSITE TOP, common in open Sierran forests, is named after one of California's pioneering botanists and geologists, William Henry Brewer.

only on about 36,000 acres primarily in the southern Sierra, and the rare tree-anemone (*Carpenteria californica*), with its "fried egg" flowers only growing in the wild just east of Fresno. Some endemics are extremely rare, such as the tiny Twisselmann's nemacladus (*Nemacladus twisselmannii*), which is known to occupy only one acre of land on earth. Even today, new plant species are still being discovered in remote and isolated sections of this mountain range.

The Sierra Nevada is prone to dry-season wildfires. Attempts in this century to limit wildfires have altered these forest ecosystems, increasing understory vegetation density and the potential for catastrophic wildfire. Land managers are reexamining the role of fire in maintaining long-term ecosystem stability.

Nearly forty percent of the Sierra Nevada is managed for its natural values as national parks, national forest wilderness areas, and wild and scenic rivers. These stunning public wilderness areas attract millions of visitors each year, which takes its toll on natural lands. Many wilderness areas now require permits, place restrictions on uses such as numbers of stock animals and foot travel, and prohibit wood burning at higher elevations to minimize damage to natural areas. Grazing and other range management practices need to be carefully monitored in meadow and riparian areas to ensure that these ecosystems are not degraded. At lower elevations the blue oak woodland and chaparral communities on the western slopes are mostly in private ownership and are being degraded at a rapid rate by urban expansion. It is hoped that local communities will plan ahead to conserve the natural character and floral diversity of the foothills.

James R. Shevock

IONE

Near the town of Ione and east of Sacramento in the foothills of the central Sierra Nevada, there is a series of island-like kaolin clay outcrops known as the Ione Formation. The ancient red and yellow soils of the Ione Formation originated under earlier tropical climates and are highly weathered and extremely acidic, high in iron and aluminum, and low in most important plant nutrients. Because few plants can tolerate these extreme and often toxic conditions, plant competi-

tion is limited. Species that have adapted to Ione Formation soils are mostly restricted to this area.

Fire plays an important role in this ecosystem. Intense heat increases seed germination of Ione manzanita (*Arctostaphylos myrtifolia*) populations. Fire also creates openings in the chaparral where rare herbaceous species such as Ione and Irish Hill buckwheats (*Eriogonum apricum* var. *apricum* and *E. apricum* var. *prostratum*) can gain a foothold.

Populations of Ione manzanita and Ione buckwheat are protected on lands owned and managed by the U.S. Bureau of Land Management and the California Department of Fish and Game (DFG). The DFG protects one of the two known populations of Irish Hill buckwheat on its Irish Hill Ecological Reserve and the largest population of Ione buckwheat on its forty-acre Apricum Hill Ecological Reserve, where the species was discovered in 1954. The Apricum Hill Ecological

Reserve is managed by the DFG to protect the rare plants, but the area is too small to ensure their long-term viability and conservation. Nearby populations on privately owned lands remain vulnerable to sand and clay mining and other human activities.

Rodney G. Myatt

One of Ione's best known endemic plants is Ione manzanita (Arctostaphylos myrtifolia), LEFT, *a sprawling dark green manzanita that dominates low hills and exposed slopes,* TOP, *and contrasts sharply with the taller gray-leaf manzanita* (Arctostaphylos viscida). *Another rare endemic, Ione buckwheat* (Eriogonum apricum *var.* apricum) *grows on soils apparently too acidic or too high in aluminum even for manzanita. • Prostrate Irish Hill buckwheat* (Eriogonum apricum *var.* prostratum), ABOVE LEFT, *occurs in small, scattered populations in openings among the manzanita. • Rock-rose* (Helianthemum suffrutescens), ABOVE, *a foothill chaparral shrub, blooms April to May.*

BLUE OAK OF THE VALLEY FOOTHILLS

In the hot, arid foothills surrounding California's Great Valley, blue oak (*Quercus douglasii*) thrives, the only oak species that tolerates midday temperatures of 100 degrees F. or more and annual rainfall of fifteen inches or less. Blue oak grows in single-species stands for over half of its over three-million-acre distribution, enduring drought and parching winds. In mixed-species stands its most common associate is gray pine (*Pinus sabiniana*).

Blue oak usually develops into a medium-sized tree, averaging thirty to forty feet, with a gently rounded canopy. As a mature tree, it becomes picturesque and craggy, evenly spaced across the landscape with each tree maintaining its own required space. In unusually hot summers, blue oaks become dormant and drop their leaves early.

Special adaptations enable the oak to thrive where it grows. Acorns germinate rapidly and early when conditions are favorable; roots grow more rapidly than those of other oaks even during cooler winter months; the tree maintains a rather small canopy of leaves to conserve energy and its leaves have several drought tolerant features.

Until the 1970s, the majority of blue oak stands were privately owned and managed as ranches. Today, many of these ranches are being subdivided for housing developments, raising urgent concerns about blue oak habitat fragmentation and loss.

Pamela C. Muick

When spring carpets of wildflowers have faded, leaves of the blue oak emerge, a halo of blue-green opalescence, BELOW. *As they age, these leaves become tough and covered with a waxy coating which conserves moisture and gives the tree its bluish cast. • The sturdy trunks of the blue oak,* BELOW LEFT, *are covered with small scales.*

PINE HILL

Not far from the urban sprawl of Sacramento in western El Dorado County, Pine Hill is located on exposed parallel bands of gabbroic and serpentine rocks that form pockets of unusual soils. Some rare plants grow here in the foothills of the Sierra Nevada and nowhere else.

The unique combination of rocks and its resulting soils support eight rare plants and one rare natural community, gabbroic northern mixed chaparral. With the exception of El Dorado bedstraw (*Galium californicum* ssp. *sierrae*), which is often found in shady oak forests, Pine Hill's rare plants all occur within the sparse chaparral community, a low-growing, shrub-dominated community that comes into full flower in early spring.

Part of Pine Hill is protected as an ecological reserve by the California Department of Fish and Game, and a portion is owned and managed by the California Department of Forestry and

The rare Pine Hill flannelbush (Fremontodendron decumbens), ABOVE, *is particularly striking with its large yellow-orange flowers. • The scientific name of Pine Hill ceanothus (Ceanothus roderickii),* RIGHT, *commemorates Wayne Roderick, longtime director of the East Bay Regional Parks Botanic Garden at Tilden Park. • Layne's ragwort (Senecio layneae),* BELOW, *is another of the eight rare plants found on Pine Hill. • Relatives of the cheery yellow El Dorado County mule ears (Wyethia reticulata),* FAR RIGHT, *have enormous fuzzy leaves that resemble mule ears. • The dark rocks scattered about on Pine Hill,* ABOVE RIGHT, *are gabbro, a dense ultramafic rock that gives local soils their special character.*

Fire Protection. Much of the rest of the unusual gabbro and serpentine habitat is unprotected and lies in the growth path of Sacramento's sprawling foothill communities. Another threat to the rare plants of Pine Hill is the illegal use of their habitat by off-road vehicles. The Department of Fish and Game is working with El Dorado County to establish a system of preserves that would protect some of the best remaining examples of the rare plant habitat of Pine Hill.

Roxanne L. Bittman

RED HILLS OF TUOLUMNE COUNTY

One of the jewels of the state's serpentine outcrops is the scenic Red Hills found near Jamestown. The hills are named for the reddish soil that often results from the weathering of green serpentine into soils that are toxic to plants.

Red Hills vegetation, restricted to these serpentine soils, is limited to diminutive grasses, a sparse cover of gray pine (*Pinus sabiniana*), chaparral shrubs, and a number of rare serpentine-tolerant plants. In contrast, the surrounding foothill soils, formed mainly from granite, support luxuriant grasses and scattered oak trees, a typical California savanna. A casual observer might think that a fire had

caused the open and barren appearance of the Red Hills, where rusty-colored oxidized soils heat up more quickly and spring comes a little earlier than on the adjacent foothill soils.

Each spring, thousands of people travel here to enjoy and photograph the spectacular profusion of wildflowers. Years of human abuse, however, have put the Red Hills at risk. Much of the area is owned and managed by the U.S. Bureau of Land Management (BLM). The agency has made numerous attempts to curb off-road vehicle activities, target shooting, and illegal dumping. In spite of detailed management plans over the past decade, the agency has been forced to close the entire area to off-road vehicles and to forbid all target shooting. The entire 7,100 acres is designated an Area of Critical Environmental Concern by BLM. Ongoing threats to the area include activation of existing mining claims, urbanization, and abuse by visitors. BLM hopes that a new public awareness campaign stressing the value of biological diversity may help to preserve the uniqueness of this ecological island.

Joseph L. Medeiros

Red Hills vegetation, OPPOSITE, *dominated by gray pine* (Pinus sabiniana) *and shrubby buck brush* (Ceanothus cuneatus)*, has a sparse otherworldly quality in dry years and seasons, but comes alive with goldfields* (Lasthenia spp.) *in wet years.* • *Rawhide Hill onion* (Allium tuolumnense)*,* OPPOSITE TOP, *is known from only twenty occurrences and is threatened by grazing and mining.* • *Bitterroot* (Lewisia rediviva)*,* TOP LEFT, *prefers rocky habitats with little soil. Its widespread abundance in California in no way diminishes its special beauty.* • *The world's only population of the blue Chinese Camp brodiaea* (Brodiaea pallida)*,* ABOVE LEFT, *lines a small section of an ephemeral creek on privately owned land.* • Navarretia pubescens*,* ABOVE, *is sticky and glandular and is found on open clay or gravel slopes.*

GIANT SEQUOIA

Giant sequoia (*Sequoiadendron giganteum*) trees are the most massive in the world today, and probably of all time. In the Calaveras North Grove, traditionally believed to be where the first giant sequoia was seen by Europeans, a tree thirty feet in diameter was cut down to make a dance floor where thirty couples could waltz.

Coast redwood (*Sequoia sempervirens*), giant sequoia's only living North American relative, is found along the coast in the Klamath and North Coast ranges south to Santa Barbara, where there is year-round moisture from summer fog and winter rain.

Giant sequoias grow with ponderosa pine (*Pinus ponderosa*), sugar pine (*P. lambertiana*), white fir (*Abies concolor*), incense cedar (*Calocedrus decurrens*), and black oak (*Quercus kelloggii*). Ranging in size from one to 4,000

acres, the isolated sequoia groves total approximately 36,000 acres. Scientists believe that giant sequoia may be more widespread today, because of increased moisture, than at any time in the past 10,000 years. John Muir apparently came to the same conclusion more than a hundred years ago after searching most of the giant tree's range for traces of extinct groves, but finding none.

About ninety percent of living giant sequoias grow on public lands and receive some level of protection. Suppression of wildfires in and near giant sequoia groves has inhibited the establishment of new trees, which are overshadowed by such shade-tolerant trees as white fir and incense cedar. Current management includes prescribed burns to reduce understory competitors to provide room and light for seedling giant sequoias.

Robert R. Rogers

TREE-ANEMONE

In horticultural circles tree-anemone (*Carpenteria californica*) is regarded as a prize, but botanists consider it a "living relic." It is an ancient species of shrub with no close relatives, the only species in the genus *Carpenteria*. Growing to a height of about eight feet, tree-anemone is most impressive in late spring when its branches are covered with large, snowy white, anemone-like flowers.

Native populations grow only in eastern Fresno County between the San Joaquin and King rivers. It inhabits granitic soils in chaparral, gray pine, and oak woodland communities between 1,500 and 4,000 feet elevation. There this beauty is often associated with Mariposa manzanita (*Arctostaphylos mariposa*), chaparral whitethorn (*Ceanothus leucodermis*), and interior live oak (*Quercus wislizenii*).

Populations on privately owned lands remain vulnerable to clearing for residential development. Fire suppression policies of the past fifty years have negatively affected the species, because natural wildfires create openings for seed germination, remove competition, and promote sprouting from the plant's base.

Three nature preserves have been established to protect some of the remaining threatened tree-anemone populations. The U.S. Forest Service has established the Carpenteria Botanical Area and the Backbone Creek Natural Area on lands near Auberry. The Nature Conservancy has established the Mary Miller Preserve on Black Mountain for educational and research purposes, as well as for protection of the tree-anemone.

John Stebbins

The tree-anemone, ABOVE, *is a shrub of the mock orange family, Philadelphaceae. Its beautiful white flowers, which appear in May and June, measure about two and a half inches across. First discovered in 1845 by John Fremont, the early California explorer, this rare evergreen was an immediate hit in royal gardens in England. Though horticulturally popular in California in native plant gardens, tree-anemone has an extremely limited distribution in the wild.*

Twenty million years ago, the genus Sequoiadendron *was more widespread. Today the giant sequoia is found in 75 groves within a narrow 260-mile belt in the Sierra Nevada at elevations between 4,500 and 7,500 feet. Here the fall colors of mountain dogwood (*Cornus nuttalli*),* OPPOSITE, RIGHT, *enliven a grove in the Sierra National Forest. • Well known to Sierran travelers, the snow plant (*Sarcodes sanguinea*),* OPPOSITE, TOP LEFT, *is so named for its reliable appearance as snows melt. Less well known is its close relationship to rhododendrons, blueberries, and other members of the family Ericaceae. • California is home to the greatest diversity of North American true lilies, represented here in the Sierra by Kelly's lily (*Lilium kelleyanum*),* OPPOSITE, BOTTOM LEFT. *• Various blazing stars (*Mentzelia spp.*),* BELOW LEFT, *occur in California. Like this one, they are usually in open, often arid habitats rather than deep within forest groves. • The lovely rosy fairy lantern (*Calochortus amoenus*),* BELOW RIGHT, *is known only from the foothills of the southern Sierra, well below the giant sequoia belt. • All members of the genus* Linanthus, *including mustang clover (*L. montanus*),* RIGHT, *are small annuals of open, often arid places; this one favors open conifer woodlands.*

135

BUTTERFLY VALLEY

Near the northern end of the Sierra Nevada in Plumas County, the Butterfly Valley Botanical Area is managed by the U.S. Forest Service as a true botanical treasure of fen, meadow, and forest. About five miles southwest of Quincy, this Sierran valley is home to more than 500 native plant species. Here at 3,700 feet, cold mountain springs find their way to the surface and flow across a sloping meadow, creating a home for the fascinating insectivorous plants, California pitcher plant (*Darlingtonia californica*) and round-leaved sundew (*Drosera rotundifolia*), and for other fen and marsh plants. These California pitcher plant fens offer a rare glimpse into the world of wild carnivorous plants and always remain attractive to both insects and botanists.

Twenty-four members of the lily family have been recorded in the vicinity of the California pitcher plant fen, including wild onion (*Allium* spp.), wild hyacinth (*Brodiaea* spp.), mariposa lily (*Calochortus* spp.), camas lily (*Camassia* spp.), fawn lily (*Erythronium* spp.), fritillary (*Fritillaria* spp.), bog asphodel (*Narthecium* spp.), fat and slim solomon seal (*Smilacina* spp.), and seven genera of orchids: coralroot (*Corallorhiza* spp.), lady's-slipper (*Cypripedium* spp.), stream orchid (*Epipactus* spp.), rattlesnake plantain (*Goodyera* spp.), bog orchid (*Platanthera* spp.), twayblade (*Listera* spp.), and ladies tresses (*Spiranthes* spp.).

The Butterfly Valley Botanical Area and the surrounding mixed coniferous forest and watershed is part of Plumas National Forest. The forests in this watershed have been logged numerous times in the past, though today logging is not permitted within botanical area boundaries. Until conflicts between private interests, environmental groups, and land management agencies are resolved, the fate of these forests and the future of the California pitcher plant fens remain insecure.

Nancy and Bill Harnach

Erroneously called cobra lily by some, the California pitcher plant (Darlingtonia californica), OPPOSITE, is not a lily, but a member of a uniquely American family, Sarraceniaceae, the New World insectivorous pitcher plants. The swollen hollow leaves of the pitcher plant fill partway with water and digestive enzymes as they grow. • When unwary insects enter the trap, light patches, FAR LEFT, confuse them, eventually causing the prey to fall into the liquid. From dead and decaying insects, the plants are able to absorb nutrients, otherwise in short supply in a fen. • The first field observations of the fen plants in Butterfly Valley, ABOVE, were probably made between 1873 and 1878 by Rebecca Merritt Austin, a keen observer and collector of plants. Many of her plant collections were sent to Edward Lee Greene, botany professor at the University of California, who named several of them as new species. • The sundew, Drosera rotundifolia, LEFT, is also found in the fens of Butterfly Valley.

137

SIERRA VALLEY

On a pleasant twenty-five mile drive north from Truckee on state Highway 89, through the pine-covered mountains of the east slope of the Sierra Nevada, one can look down nearly 1,000 feet to where a large mountain valley about the size of Lake Tahoe sprawls across an ancient lake bed—the Sierra Valley, a land of harsh extremes, breathtaking beauty, and vast biological diversity.

Tucked in against the eastern edge of the Sierra, the average elevation of the Sierra Valley floor is 4,950 feet, with the encircling mountains in some areas climbing steeply to more than 8,000 feet. The climate is typical of the Great Basin. Moisture comes usually in the form of annual snowfall, up to eight feet on occasion but more commonly about three feet. On the east side of the valley, no more than ten miles away, moisture rapidly diminishes to about half that. Following a snowfall, the temperature often drops to 20 to 30 degrees below zero. Summers bring the other end of the spectrum, dry weather with temperatures into the 90s.

It is the extreme variation in daytime temperatures, which can amount to 40 or 50 degrees, that make the valley suited to only the heartiest of plants. In this environment, spring bursts forth with a spectacular, if diminutive, floristic world. Dry, rocky, high desert lands to the east, shallow depressions with small vernal pools to the west, mountain streams cascading

A stunning sea of blue camas lilies (Camassia quamash), ABOVE AND OPPOSITE TOP LEFT, covers many of the wetter meadowlands on the southern end of the valley floor. Blue camas was an important food source for Native Americans wherever it grew in California; the bulbs were roasted. • One of Sierra Valley's many rare plants is Ivesia aperta var. aperta, LEFT, also known as the Sierra Valley ivesia. Its flowers are visited by a number of different insects, including ants. • Leopard lily (Lilium pardalinum), RIGHT, grows three to six feet tall, often in large groups along moist stream-banks at lower elevations. • Bitter cherry (Prunus emarginata), OPPOSITE TOP RIGHT, is named for the small fruits, which are intensely bitter.

from high peaks to the more moist valley floor, springs oozing from the base of the mountains—it is the variety of habitats and climatic extremes that makes Sierra Valley a springtime paradise as well as a home for both the common and the rare.

No rare species of plants are formally protected at present, and most of Sierra Valley is privately owned. The U.S. Forest Service has developed guidelines for the preservation of *Ivesia* species on federal lands. These guidelines are shared with interested private landowners.

Nancy and Bill Harnach

TAHOE BASIN

The deep indigo blue water of Lake Tahoe can be seen from any point in the Tahoe Basin and from most of the distant peaks. Mountains encircling this glacier-carved lake basin in the northern Sierra Nevada reach toward 11,000 feet and seldom drop below 8,000 feet. They lie buried in snow from December to April. From June through August, however, when waterfalls and streams leap over granitic boulders and wind their way through alpine meadows, the mountains come alive with the fragrance and color of a glorious succession of wildflowers.

With a mixture of granitic, volcanic, and metamorphic rock forms that have decomposed to create a variety of soil types and plant habitats, Tahoe becomes a meeting place for plants from both the western flanks of the Sierra Nevada and the eastern Great Basin. Between 6,400 and 9,000 feet elevation, the red fir forest comes into its own with its common associates, the handsome and fragrant Jeffrey pine (*Pinus jeffreyi*), western white pine (*P. monticola*), lodgepole pine (*P. contorta* var. *murrayana*), mountain hemlock (*Tsuga mertensiana*), and western juniper (*Juniperus occidentalis*). Here in the understory are often found the lovely soft rose-colored flowers of *Spiraea densiflora*, western serviceberry (*Amelanchier pallida*), and occasional thickets of snow brush (*Ceanothus cordulatus*).

The summer beauty of the national forests in Tahoe Basin brings thousands of visitors who explore the lakes, meadows, and forests. Unfortunately, the impact of such high use is often apparent in eroded trails and, in some places, invasions of exotic plants that gain a foothold in disturbed soils.

Julie Carville

California red fir (Abies magnifica *ssp.* magnifica), RIGHT, *encircles Emerald Bay, Lake Tahoe, just above the water level, where it shares the forest mostly with lodgepole pine* (Pinus contorta *ssp.* muricata). *Higher up, as shown here, it mingles with other upper montane conifers, but none aspires to the magnificence of the largest red firs, some reaching 200 feet in height.* • *The Tahoe Basin's numerous wet meadows grow whimsical elephant heads* (Pedicularis groenlandica), BELOW LEFT, *looking like tiny pink elephants with rounded foreheads and upraised trunks.* • *Dwarf lupine* (Lupinus lepidus *var.* sellulus), BELOW RIGHT, *forms low patches of color on the forest floor when in bloom in mid-summer.* • *Pink heather* (Phyllodoce breweri), OPPOSITE, ABOVE LEFT, *named in honor of a Greek sea nymph, is one of the best known and loved of the Sierran plants. It forms mats on acid soils around high-elevation lakes, particularly around moist rocky surfaces.* • Aster ascendens, OPPOSITE, ABOVE MIDDLE, *scarce in the Tahoe Basin and more common in areas with Great Basin influence, is found in meadows and red fir forests.* • *Sneezeweed* (Helenium bigelovii), OPPOSITE, ABOVE RIGHT, *brightens wet meadows in mid- to late summer.*

LAKE WINNEMUCCA

During summer months, with a short hike of about three miles and an elevation gain of less than a thousand feet, one can explore the breathtaking beauty of the alpine flora of the Sierra Nevada found at Lake Winnemucca and Frog Lake near Carson Pass (8,500 feet) in the Eldorado National Forest. In all of the Sierra, there are few places where the alpine flora is as diverse and accessible. Along a trail to the lakes are found more than 450 native plants, none rare or endemic, but growing in endless little meadow, forest, and alpine landscapes of wondrous beauty and only paces apart.

A key to the diversity of vegetation at Carson Pass is geologic: hard granite, porous volcanics, crumbly metamorphics all appear here in a mixed-up puzzle. Topography and water provided by snowmelt, springs, and seeps create a quilt of varied plant communities and colorful wildflowers.

The unique feature of the Carson

Red giant paintbrush (Castilleja miniata), OPPOSITE, *and yellow arrowhead butterweed* (Senecio triangularis) *share space with blue lupine and dozens of other wildflowers at Lake Winnemucca.* • *Mountain mule ears* (Wyethia mollis), OPPOSITE BOTTOM, *frame a view of Winnemucca Lake and Round Top in the Mokelumne Wilderness.* • *The lovely crimson columbine* (Aquilegia formosa), TOP, *attracts an assortment of hummingbirds and bumblebees to the generously supplied nectar of its flowers. At the uppermost extremes of its range it is replaced by a white, moth-pollinated relative* (A. pubescens) *called Coville's columbine.* • *Monument plant* (Swertia radiata), ABOVE, *is a greenish representative of the gentian family, Gentianaceae. Like the agaves, it is a perennial that flowers only once and dies shortly thereafter.* • *Like the paintbrushes, the azure penstemon* (Penstemon azureus), BELOW, *is beautiful, represents a group that is diverse in California, and is a member of the Scrophulariaceae or snapdragon family.*

Pass flora is the juxtaposition of diverse plant communities spanning an array of subalpine and alpine sites. At the junction of three botanical provinces, lush meadows are found with eight-foot-tall, many-branched larkspur (*Delphinium polycladon*) and brightly colored combinations of red-flowered giant paintbrush (*Castilleja miniata*), yellow-flowered arrowhead butterweed (*Senecio triangularis*), and blue-flowered large-leaved lupine (*Lupinus polyphyllus*). Yards away, alpine vegetation occurs with diminutive cushion plants such as dense-leaved draba (*Draba densifolia*) or southern Sierra chaenactis (*Chaenactis alpigena*) only a few inches tall. Subalpine forests of whitebark pine (*Pinus albicaulis*) and hemlock (*Tsuga mertensiana*) are bordered by slopes covered with four fragrant sagebrush species (*Artemisia* ssp.).

The U.S. Forest Service has designated this as the Roundtop Botanical Area. There is a kiosk at the trailhead on Carson Pass, built and staffed by volunteers during summer months. Because of easy access and the large number of summer visitors, good trail manners are particularly important for long-term safety of the plants.

Dean Taylor

TRUMBULL PEAK

Trumbull Peak overlooks the Merced River and across the Stanislaus National Forest at Half Dome in Yosemite Valley. The shallow, rocky soils of this scenic, exposed ridgetop composed of ancient metasedimentary rock are home to three rare plants: Yosemite onion (*Allium yosemitense*), Congdon's woolly sunflower (*Eriophyllum congdonii*), and Congdon's lewisia (*Lewisia congdonii*). Throughout their limited ranges, these rare plants occur only on shallow rocky soils surrounded by manzanita (*Arctostaphylos* spp.) and chamise (*Adenostoma fasciculatum*) chaparral; black oak and ponderosa pine forests cover the north slopes.

Trumbull Peak is gaining in popularity with both wildflower enthusiasts and history buffs, and visitors also seek the peak's spectacular views. More than 100 native plant species can be seen along the access road and trail. Remnants of a rail and cable system that transported logs to the Merced River in the 1920s remain in the area.

Trumbull Peak is now designated by the U.S. Forest Service as a botanical and historical special interest area. The Forest Service will be increasingly challenged to accommodate visitor use of Trumbull Peak while protecting its rare plants and historic features.

Jennie Haas

Congdon's lewisia (Lewisia congdonii), ABOVE RIGHT, *a perennial herb, has a wider range than its rare Trumbull Peak companions, but it is found only on north-facing cliffs.* • *The perennial, pink-flowered Yosemite onion* (Allium yosemitense), ABOVE, *has only one known occurrence outside the Merced watershed.* • *Across the Merced River Valley from Trumbull Peak, the famous half-dome rocks,* RIGHT, *catch the afternoon light.*

JEFFREY PINE

On the steep eastern face of the Sierra Nevada, the magnificent Jeffrey pine (*Pinus jeffreyi*) reigns from Mt. Shasta in the north to the San Jacinto and San Bernardino mountains in the south. Elsewhere in the Sierra at high elevations in upper montane and subalpine conifer forests, this excellent timber tree replaces its close cousin, ponderosa pine (*P. ponderosa*). But Jeffrey pine's versatility is best expressed in its strong preference for serpentine-derived soils, which extends its elevational range nearly to sea level in the North Coast and Klamath ranges and to below 2,000 feet in the Sierra.

Trees may grow to 175 feet, but it is the architecture of the tree, the stunning reddish bark that divides into glossy hard plates, the tufts of dark green needles borne on slender, slightly upturned branches, that make Jeffrey pine such a picturesque part of the California landscape. At high elevations where the trees are exposed, as on Sentinel Dome in Yosemite, Jeffrey pine sprawls along the ground or lies flat on rocks in true alpine growth form. At lower elevations, it grows tall and straight in park-like forests, with a longer, often broader crown than the ponderosa's.

Jeffrey pine sometimes intermingles with ponderosa pine, especially in the Sierra and Southern California's high mountains; hybrids are common in these zones. When identifying a yellow or Jeffrey pine by its cone, turned-in prickles immediately signify the latter. The bark of the Jeffrey pine has a recognizable fragrance, for some similar to rotting apples, for others, to pineapple or vanilla.

The cones are the largest pine cones in the west, next to the gigantic cones of the grey, Coulter, and Torrey pines. At the moonscape serpentine barrens near New Idria on San Benito Mountain, digger, Coulter, and Jeffrey pines occur together, a mecca for lovers of huge-coned pines.

Mark W. Skinner

The Jeffrey pine was named for Scottish botanical explorer John Jeffrey, who first collected it in Shasta Valley in 1852. At lower elevations it grows in park-like stands, as shown at the Domeland Wilderness Area, ABOVE, *and at Church Dome,* BELOW, *both in Sequoia National Forest in Tulare County.*

ALONG THE JOHN MUIR TRAIL

The splendor of the Sierra Nevada is punctuated by its breathtaking "high country," a lofty zone above 10,000 feet in elevation that runs more than 150 miles from Sonora Pass (9,626 feet) southward to Olancha Peak (12,130 feet). Within this Sierran high country lies one of the most famous of American trails, the John Muir Trail.

Begun in 1915, the year following the death of John Muir, the trail was designed by early members of the Sierra Club who had collectively explored much of the high Sierra. By 1938, the trail was completed, opening an adventurous route between Yosemite and Sequoia national parks.

The John Muir Trail begins below Vernal Falls at Happy Isles in Yosemite, 4,035 feet above sea level. California black oak (*Quercus kelloggii*), incense cedar (*Calocedrus decurrens*), and ponderosa pine (*Pinus ponderosa*) join other species of trees and shrubs in a typical mid-elevation Sierra forest. Before the wanderer has hiked five miles, more than two thousand feet of elevation has been gained to find water from boisterous Vernal and Nevada falls cascading down over sheer rock faces. Behind the famed granite monuments of Half Dome and Clouds Rest, the trail continues past Jeffrey pine (*Pinus jeffreyi*) and white fir (*Abies concolor*) upward to Tuolumne Meadows at 8,500 feet in the high Sierra.

To the south the trail gains elevation, taking trekkers through passes such as Donohue (11,050 feet), Muir (11,955 feet), Mather (12,080 feet), and Forester (13,200 feet). These are the haunts of whitebark pine (*Pinus albicaulis*) and mountain hemlock (*Tsu-*

ga mertensiana). The southern terminus of this magical route crosses the summit of Mt. Whitney at 14,496 feet, the highest place in North America outside Alaska.

Watching the sunset from here lulls one into complacency regarding protection of this mountain paradise. Yet, even the high Sierra is within the reach of the millions who live below. While the herds of sheep that once ravaged alpine meadows are gone, they have been replaced by thousands of footprints of humans, horses, mules, and even llamas. Cattle grazing permits are still issued for some high-country meadows and the pollution cloud from cars winds it way up the canyons into the alpine air. Securing this national treasure will require sensitive and grand efforts by us all.

Joseph L. Medeiros

Discovering icy-cold natural lakes along the John Muir Trail with surrounding gardens of wildflowers, ABOVE, *is one of life's great treats.* • *The dome-shaped flowerheads of sneezeweed* (Helenium bigelovii), OPPOSITE, TOP LEFT, *are common at mid-elevations in Sierran meadows.* • *Often forests or meadows are carpeted with long-stemmed clover* (Trifolium longipes), OPPOSITE, BOTTOM LEFT. • *Flowers of American bistort* (Polygonum bistortoides), OPPOSITE, RIGHT, *in the buckwheat family are conspicuous in many mountain meadows. It is sometimes irreverently called dirty socks.* • *The early-blooming and common* Fritillaria atropurpurea, LEFT, TOP, *found in leaf mold under trees, is easily confused with the uncommon* F. pinetorum, *found on shaded granitic slopes.* • Helianthus annuus, LEFT, BOTTOM, *is a variable member of the sunflower family that readily hybridizes with other species.*

147

PIUTE RANGE

Within the extreme southern extension of the Sierra Nevada in Kern County, the remote Piute Range forms a complex geologic island of diverse rock types on which grow a wealth of plant species. The Piute Range is arid in comparison to the adjacent Greenhorn and Breckenridge portions of the southern Sierra Nevada. Vegetation on the Piutes forms a complex of woodland, chaparral, conifer forests, and the world's largest grove of the rare Piute cypress (*Cupressus arizonica* ssp. *nevadensis*).

The Piute Range's conifers grow in open, widely spaced stands because of limited precipitation. Pinyon pine (*Pinus monophylla*) and Joshua tree (*Yucca brevifolia*) are common on the eastern slope of the Piutes, along with extensive stands of blackbush (*Coleogyne ramosissima*).

Marble outcrops along the eastern and northern slopes of the Piutes support species more commonly found on the carbonate rocks of the Mojave Desert. One of the most interesting plants on these marbles is Nevada greasewood (*Glossopetalon spinescens*), a spiny shrub that represents the only occurrence of the crossosoma family in the Sierra Nevada.

The southern Sierra Nevada has suffered impacts from varied human activities. Early in this century the Piutes were a source of lumber for small communities in the Lake Isabella Valley. Gold mining is still common in the Piutes. The area is accessible from California Highway 178 near Lake Isabella, largely on graded dirt roads or four-wheel drive routes.

The Piute Range is administered by the U.S. Forest Service and the Bureau of Land Management, with a few scattered private parcels within the federal boundaries. Because of widespread mining in the Piutes prior to the establishment of Sequoia National Forest, few acres have wilderness-quality

Cottonwoods (Populus fremontii) *glow in Kern River Canyon,* LEFT, *accentuating arid slopes so typical of the Piute region.* • *Rocky crevices in the Piute high country support Hall's daisy* (Erigeron aequifolius), BELOW. • *Good eyesight and a sturdy back are required to appreciate the delicate beauty of the tiny calico monkeyflower* (Mimulus pictus), OPPOSITE, TOP LEFT, *known only from the Greenhorn Mountains to the northwest of the Piute Range.* • *The Piutes have been recognized for their botanical virtues since the last century. Explorers such as German botanist Carl A. Purpus made numerous plant collections here in the 1890s, including the restricted Kern County larkspur* (Delphinium purpusii), OPPOSITE, TOP RIGHT, *and others both new to science and endemic to the Piute region.* • *The Mojavean floristic element, here represented by Joshua trees* (Yucca brevifolia), RIGHT, *is but one major floristic influence promoting botanical richness in the Piutes.*

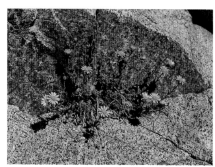

landscapes consistent with federal designations. Two botanical areas, where plants receive special protection, have been established. The Bodfish Piute Cypress Botanical and Natural Area, near the town of Bodfish, includes the largest grove of Piute cypress. The other botanical area is Inspiration Point, a dolomitic limestone outcrop with windswept limber pine (*Pinus flexilis*), mountain-mahogany (*Cercocarpus ledifolius*), and dwarf mountain maple (*Acer glabrum* var. *torreyi*).

James R. Shevock

OWENS PEAK AND SPANISH NEEDLE

Along the rugged, arid eastern crest of the southern Sierra Nevada in Kern County, an assemblage of plants and plant communities comes together from distinct and diverse sources. Floristic elements from the Sierra Nevada, the Great Basin, Mojave Desert, and Transverse Ranges of Southern California overlap at Owens Peak and Spanish Needle. Dominated by granitics, the Owens Peak and Spanish Needle area also contains marbles and metamorphic rock types, all contributing to specialized local soils and plant habitats, and these in turn, to the presence of many rare species.

The northeast slope of 8,543-foot Owens Peak is a complex conifer woodland of Jeffrey pine (*Pinus jeffreyi*), sugar pine (*P. lambertiana*), limber pine (*P. flexilis*), western juniper (*Juniperus occidentalis*), white fir (*Abies concolor*), pinyon pine (*P. edulis*), mountain-mahogany (*Cercocarpus betuloides*), dwarf mountain maple (*Acer glabrum*), and one of the largest stands in the Sierra of Parry beargrass (*Nolina parryi*), a yucca-like plant.

Three rare and endemic plants have recently been discovered and named from this remote and, until recently, little explored area. Owens Peak lomatium (*Lomatium shevockii*), a member of the carrot family, is known from two populations, one on Owens Peak and another on nearby Jenkins Peak. Sweet-smelling monardella (*Monardella beneolens*) grows among rocks on

Owens Peak, LEFT, *is but one of several dramatic peaks signalling the southern terminus of the Sierra.* • *Parry beargrass* (Nolina parryi), ABOVE LEFT, *dots the granitic face of Owens Peak with flower stalks up to eight feet tall and inflorescences of cream-colored flowers three feet in diameter.* • *The entire population of Spanish Needle onion* (Allium shevockii), TOP RIGHT, *is bisected by the Pacific Crest Trail, and appears to reproduce only vegetatively.* • *The freckled milk-vetch is represented by nearly twenty varieties in California alone, including this one* (Astragalus lentiginosus *var.* kernensis), ABOVE RIGHT, *which is endemic to the Kern Plateau.* • *The comical Kelso Creek monkeyflower* (Mimulus shevockii), BELOW LEFT, *is one of California's significant botanical discoveries of recent years.* • *Buckwheats are phenomenally diverse in California, and are represented by about 200 different kinds. Many, like the Needles buckwheat* (Eriogonum breedlovei *var.* shevockii), BELOW RIGHT, *are endemic to small areas.*

the peak, where its glandular leaves emit a sweet minty fragrance. Spanish Needle onion (*Allium shevockii*) is restricted to only one location in the world.

The Owens Peak and Spanish Needle area is owned and managed by the Bureau of Land Management (BLM), which maintains a sensitive plant species management program. Although there are some mining and off-road vehicle uses on the desert floor at the base of these peaks, the upper slopes are remote and pristine, and are proposed for wilderness designation by BLM. Access to the Owens

Peak and Spanish Needle area is provided by the Pacific Crest Trail from the south at Walker Pass, from the north via Lamont Meadows, and from the east by a steep, unmaintained trail.

James R. Shevock

CHANNEL ISLANDS

California's Channel Islands comprise the eight northernmost islands of an archipelago that extends from Point Conception down the coast into Baja California. Nonetheless, each of the islands serves, in its own unique way, as a vitally important refuge for plant species that have disappeared or have become imperiled on the mainland.

The Channel Islands in California fall into two quite distinct groups: the Northern Channel Islands (San Miguel, Santa Rosa, Santa Cruz, and Anacapa), which appear to be erosional remnants of the mainland Transverse Ranges such as the Santa Monica and San Gabriel mountains; and the Southern Channel Islands (Santa Barbara, San Nicholas, Santa Catalina, and San Clemente), which have more diverse origins with less obvious ties to the mainland.

Because the islands are far enough from the coastline to prevent a continual interchange of seeds and other propagules with the mainland, the flora of the Channel Islands includes many rare species. Because of isolation and diverse topography, over time new species have evolved and other species that have died out on the mainland persist on the islands as relics. But the

factors that probably contribute most to the large number of paleo-endemic species found on these islands are the conditions of mild climate and frequent marine fogs, a reminder of a warmer, moister era. Santa Cruz Island ironwood (*Lyonothamnus floribundus* ssp. *asplenifolius*), for example, today grows on several of the Channel Islands but exists only on the mainland as fossils, indicating a more extensive range in the past.

Introduced grazing animals that have become feral on the islands and aggressive non-native plants introduced over the years threaten the native flora of the islands. At least twenty-eight plant species found only on the Northern Channel Islands are thus endangered throughout their entire current range. The islands display the extent of community disintegration that results from feral animals and exotic plants. At the same time, as development continues on the adjacent mainland, the role of these islands as refuges for native coastal plants will become increasingly important.

Tom Oberbauer and Steve Junak

Candleholder dudleya (Dudleya candelabrum), ABOVE LEFT, has become reestablished over much of its historic range following removal of most feral herbivores from Santa Cruz Island (pigs remain a serious problem). • Channel Island tree poppy (Dendromecon harfordii), ABOVE RIGHT, is a common chaparral component on Santa Rosa and Santa Cruz islands. It is rare on Santa Catalina Island, and was extirpated from San Clemente Island by goats and other herbivores. • The small succulent (Dudleya blochmaniae ssp. insularis), BELOW, occurs in one small area on Santa Rosa Island and nowhere else in the world. • Catalina Island, OPPOSITE, is one of five of the eight Channel Islands that are now within the Channel Islands National Park. Land ownership and degrees of resource protection vary from island to island.

NORTHERN CHANNEL ISLANDS

San Miguel and Santa Rosa islands are composed largely of eroded, ancient sedimentary rocks. They have relatively low, exposed profiles capped by broad, subdued peaks. Santa Cruz and Anacapa islands are dominated by more resistant volcanic rocks and have a more rugged topography with sharp peaks.

The ecosystems of all four northern islands have been damaged by overgrazing from sheep, which not only has resulted in severe erosion but has radically altered the composition and distribution of the vegetation. Many native plants have been almost exterminated, and the islands are completely denuded in some areas. All grazing animals were removed in the 1940s and 1950s from San Miguel (owned by the U.S. Navy) and Anacapa; now both are managed by the National Park Service.

Santa Rosa Island, previously a private cattle ranch, was acquired by the National Park Service in 1986. Under an agreement with the previous owners, ranching will continue there until 2011. The Park Service has recently removed the last of a feral pig popula-

Island oak (Quercus tomentella) *shown growing on Santa Rosa Island,* LEFT, *is the rarest of California's tree oaks.* • *Soft-leaved Indian paintbrush* (Castilleja mollis), TOP LEFT, *is an exceptionally beautiful perennial found only on San Miguel and Santa Rosa islands, where it is currently threatened by browsing and trampling.* • *Some island plants have made remarkable recoveries in recent years, most notably Santa Cruz Island bird's-foot trefoil* (Lotus argophyllus *var.* niveus), TOP RIGHT, *which has spread rapidly since the removal of feral sheep on The Nature Conservancy's property.* • *The four Northern Channel Islands—San Miguel,* ABOVE, *Santa Rosa, Santa Cruz, and Anacapa—are the exposed portions of an eighty-mile-long seamount. They differ dramatically from one another in size (from one to ninety-six square miles), in distance from the mainland (from thirteen to twenty-seven miles), in maximum elevation (from 830 to 2,470 feet), and in geology, topography, climate, and current land use.* • *Wallace's nightshade* (Solanum wallacei), BELOW, *is a rare member of the potato genus, the largest among flowering plant genera with 2,000 species.*

tion, but non-native mule deer, elk, horses, and cattle still remain. Six plants, varying in size from the majestic Santa Rosa Island Torrey pine (*Pinus torreyana* ssp. *insularis*) to the diminutive succulent Santa Rosa Island dudleya (*Dudleya blochmaniae* ssp. *insularis*), are restricted to this island.

Ninety percent of Santa Cruz Island is now owned by The Nature Conservancy, and the last ten percent, to the east, is being purchased by the National Park Service from private owners. Feral pigs, horses, and a few cattle all remain on the island. Feral sheep have been successfully removed from The Nature Conservancy's portion of land but are still present on the eastern end. The very rare Hoffmann's rock cress (*Arabis hoffmannii*) was rediscovered on Santa Cruz Island in 1985; however, the Santa Cruz Island monkeyflower (*Mimulus brandegei*) has not been seen since 1932 and is now thought to be extinct.

Steve Junak

155

SOUTHERN CHANNEL ISLANDS

Santa Catalina, some eighteen miles south of Long Beach, is the best known of the Channel Islands. For many years it was owned and developed by the Wrigley family, and the ferry connecting the town of Avalon to the mainland made this the only island easily accessible to the public. The majority of the island, now owned and managed by The Catalina Conservancy, has not been developed; large areas of chaparral, coastal sage scrub, ironwood groves, island cherry woodland, grassland, and oak woodland remain.

The native vegetation has suffered greatly from imported game animals: mule deer, wild pigs, goats, even bison.

Nevertheless, several plants endemic to this island are still found growing here: Santa Catalina Island manzanita (*Arctostaphylos catalinae*), Catalina Island mountain-mahogany (*Cercocarpus traskiae*), a subspecies of Catalina ironwood (*Lyonothamnus floribundus* ssp. *floribundus*), and Santa Catalina Island buckwheat (*Eriogonum giganteum* var. *giganteum*).

The U.S. Navy owns San Nicholas Island, where it maintains an active military base. San Nicholas is famous as the setting of the book *Island of the Blue Dolphins*. The island forms a low mesa, roughly football-shaped and consisting of erodible sandstones, made interesting because of the calcium carbonate deposits (caliche casts) that form on fossil roots. Vegetation on the island has suffered from decades of sheep grazing, but is now making a comeback following a sheep removal program.

San Clemente Island, a long, nar-

row volcanic island, rises gently from its western side to nearly 2,000 feet and then drops steeply to the sea along its eastern coast. Large canyons cut down the sides of the volcano. The island was probably once covered with coastal sage scrub and canyon woodlands of Catalina ironwood, coast live oak (*Quercus agrifolia*), and Catalina cherry (*Prunus ilicifolia* ssp. *lyonii*), but all this has been converted largely to eroded non-native grasslands by feral goats and pigs. The U.S. Navy, which also owns this island, has recently been removing these animals to allow vegetation to recover. Thirteen endemics, a relatively large number, are found on San Clemente Island and nowhere else on earth. Among these are the one- to three-foot tall, white-flowered San Clemente Island bush mallow (*Malacothamnus clementinus*) and two bulbiferous species: San Clemente Island brodiaea (*Brodiaea kinkiensis*), occurring on clay soils, and San Clemente Island triteleia (*Triteleia clementina*), found on steep canyon slopes and leeward grassy terraces.

Santa Barbara Island is one of the smallest of the Channel Islands, only one square mile in size. Owned by the National Park Service, it was once

SANTA CATALINA ISLAND MOUNTAIN-MAHOGANY

The only remaining population of Santa Catalina Island mountain-mahogany (*Cercocarpus traskiae*) is currently found in a small canyon on Santa Catalina Island. This small tree, growing to no more than twenty-six feet tall, is a relict species once much more widely distributed. When the species was discovered on Santa Catalina Island in 1897, the population consisted of forty to fifty mature trees. By 1990, only seven trees remained because of browsing and grazing by feral deer, goats, and pigs. The Catalina Island Conservancy and California Department of Fish and Game are cooperating to protect

the Catalina Island mountain-mahogany by fencing and establishing new populations.

Richard Lis

Though we do not know how extensive its range once was, Santa Catalina Island mountain-mahogany's closest relatives now survive in southern Mexico. This remaining population is threatened by loss of genetic diversity and by hybridization with the birch-leaf mountain-mahogany (C. betuloides var. betuloides), a species widespread on the island and the mainland. Surviving trees are fenced from grazing to promote seedling establishment and survival.

known for a forest of eight- to ten-foot tall giant coreopsis (*Coreopsis gigantea*), now completely destroyed on the island by rabbits introduced less than fifty years ago.

Thomas Oberbauer

Island mallow (Lavatera assurgentiflora), FAR LEFT, *is widely cultivated for its beauty.* • *San Nicolas Island buckwheat (Eriogonum grande var. timorum), OPPOSITE, TOP RIGHT, is considered endangered by the State of California.* • *Since the rabbits were removed from Santa Barbara Island in the late 1970s, the rare Trask's milk-vetch (Astragalus traskiae), OPPOSITE, LOWER RIGHT, has been rediscovered, along with endemic Santa Barbara Island dudleya (Dudleya traskiae) and Santa Barbara Island buckwheat (Eriogonum giganteum var. compactum).* • *Once believed to have been extirpated by feral animals, five living plants of Dudleya traskiae, ABOVE LEFT, were rediscovered in 1975 on Santa Barbara Island where a small population lives precariously today.* • *The four Southern Channel Islands, San Clemente, LEFT, Santa Barbara, San Nicholas, and Santa Catalina, are geologically distinct from both the Northern Channel Islands and from the mainland. Like much smaller Santa Barbara Island, San Clemente Island is primarily volcanic in origin.*

TRANSVERSE RANGES

The Transverse Ranges separate the Coast Ranges and Mojave Desert to the north from the Peninsular Ranges and Colorado Desert to the south. With one end in the Pacific Ocean and the other in the desert, the Transverse Ranges are understandably rich in plant species and communities. Twenty of twenty-nine natural communities identified in Munz and Keck's *A California Flora* occur in the Transverse Range Province, within which approximately 2,500 native plant species occur, more than half the total for the entire state.

From west to east, the Transverse Ranges include the Santa Ynez Mountains north of Santa Barbara, the mountains of central Ventura County, the Santa Monica, San Gabriel, and the San Bernardino mountains, and the Little San Bernardino Mountains and other Eastern Boundary Mountains. The Northern Channel Islands are actually an offshore extension of the Santa Monica Mountains; the Los Angeles Basin, including the Palos Verdes Peninsula, also may be included within the Santa Monica Mountains Province.

The Transverse Ranges include the greatest variety of rock types of any geologic province in California, and these range in age from late Pliocene, 1.8 million years ago, to early Pre-Cambrian, 1.8 billion years ago. Although local geology usually is complex and diverse, the western Transverse Ranges consist largely of younger, folded sedimentary rocks (Cenozoic), while the eastern ranges are mostly older, uplifted granitic fault blocks (Mesozoic).

Plant communities of the Transverse Ranges range from alpine vegetation on the summit of "grayback," as Mount San Gorgonio is locally known, through forest woodland, scrub, grassland, and marsh, down to the sea. Chaparral communities form the most extensive vegetation type in the mountains proper, covering more than fifty percent of the area, from elevations near sea level to more than 9,000 feet. Development of chaparral is favored by the prevalent steep, eroded terrain with little true soil. Coastal sage scrub and California black walnut woodland reach their best development in the southern foothills of the Transverse Ranges. Vegetation varies with elevation, soils, and distance from the sea. Mild year-round maritime temperature regimes prevail at the western seaward end of the Transverse Ranges, but these give way to wide daily and seasonal fluctuations on the desert edges.

Nearly 100 rare and endangered plants occur within the Transverse Ranges. Notable concentrations may be found near Point Conception and Point Arguello through the western Santa Monica Mountains, and in the San Bernardino Mountains near Big Bear Lake. In this basin high elevations, unusual soil types, and geographic isolation have combined to create many unique habitats such as the "pebble plains."

While many rare species and communities within the Transverse Ranges

are afforded some protection on military lands, in state parks, on U.S. Forest Service lands, or in nature reserves, their proximity to millions of people in urbanized Southern California continues to challenge management agencies. Species and habitats outside these areas remain vulnerable. Plants in coast-facing foothills, especially those adjacent to the Los Angeles Basin and San Fernando Valley, have been particularly diminished by urbanization, and several have become extinct. Plant communities that have suffered the greatest degradation and loss include coastal sage scrub, woodlands of black walnut or oak, and wetlands, including vernal pools.

J. Robert Haller

Munz's yellow buckwheat (Eriogonum umbellatum *var.* munzii), FAR LEFT, *brightens open rocky areas on Mount Pinos. • Chaparral,* ABOVE, *dominates the vegetation of the Transverse Ranges. This particular type features red shank* (Adenostoma sparsifolium)*, an important California shrub in the rose family* (Rosaceae)*. • The Transverse Ranges,* BELOW, *the only major east-west trending mountains in California, extend from Point Arguello on the mainland and San Miguel Island offshore some 325 miles to the east through Joshua Tree National Monument in the desert. Their maximum width is about sixty miles.*

FIGUEROA MOUNTAIN

In a good spring with just the right rainfall, temperature, and timing, a forty-eight-mile loop road northeast of the little town of Los Olivos in the Cachuma Valley leads to one of the best displays of wildflowers in Santa Barbara at Figueroa Mountain. The mountain summit is over 4,500 feet, providing magnificent views over the Pacific Ocean, into the back country, and over carpets of wildflowers.

Topography, past grazing and fire histories, and an assortment of soil types all contribute to a mosaic of chaparral, coastal sage, grassland, woodland, and forest communities. Gray pine (*Pinus sabiniana*), Coulter pine (*P. coulteri*), and yellow or ponderosa pine (*P. ponderosa*), big-cone spruce (*Pseudotsuga macrocarpa*), big-leaf maple (*Acer macrophyllum*), coast live oak (*Quercus agrifolia*), blue oak (*Q. douglasii*), and canyon live oak (*Q. chrysolepis*) each dominates its own special habitat, as do the manzanitas, scrub oaks, and yuccas.

The rarest plants of this area are the cream-colored Santa Barbara jewel-flower (*Caulanthus amplexicaulis* var. *barbarae*), white South Coast Range morning-glory (*Calystegia collina* ssp. *venusta*, and the pink or blue small-flowered morning-glory (*Convolvulus simulans*).

Ralph Philbrick

Owl's-clover (Castilleja exserta *ssp.* exserta), ABOVE, *adds to the brilliant spring wildflower palette at Figueroa Mountain.* • *This gorgeous wild onion* (Allium haematochiton), BELOW, *derives its latin name from its reddish bulb coat (*haemato = blood red; chiton = skin*).* • *In spring, California poppies* (Eschscholzia californica), RIGHT, *define the Golden State as do its hillsides in fall.* • *Most spectacular of all,* FAR RIGHT, *are the south-facing blue and orange fields of sky lupine* (Lupinus bicolor) *and California poppy at the summit—pure stands, mixed stands, a fragrant blaze of color.*

California's most spectacular member of the poppy family, the Matilija poppy (Romneya coulteri), LEFT, grows from the Sespe Creek drainage in the Ventura County portion of condor country south to Temescal Canyon and into Baja California. This poppy shares its name with Matilija Canyon north of the town of Ojai, where the beautiful plants are said to protect the grave of the daughter of the Matilija Indian tribe's chief.

Individual Matilija poppy flowers are some of the largest of any California native plant. Each flower has six or so crinkled white petals, which fall individually to the ground, and a center cluster of many separate yellow stamens. The tall whip-like stems that produce an abundance of flowers are sparsely clothed with irregularly shaped gray-green leaves and sprout from vigorous creeping rootstocks.

SANTA MONICA MOUNTAINS

The Santa Monica Mountains bisect the Los Angeles Basin and are dominated by steep ridges, rugged canyons, and chaparral-clad slopes, all of which are periodically swept by intense wildfires. The highest point (3,111 feet) is at the west end of the range at Sandstone Peak. From this peak, spectacular views of six of the eight Channel Islands and the southern Los Padres National Forest can be seen on clear days.

More than 100 native plant species reach their distributional limits in the Santa Monica Mountains. Bitterroot (*Lewisia rediviva*), named for the explorer, Meriwether Lewis, reaches its southern limit here. The aromatic

southern mountain misery (*Chamaebatia australis*), here at its northern limit, extends south to northern Baja California, Mexico. Several rare and endemic species grow on the volcanic rock outcrops and soils in the western part of the Santa Monica Mountains near Conejo grade. These include four species of succulent live-forevers (*Dud-*

leya spp.), the rare Conejo buckwheat (*Eriogonum crocatum*), and Lyon's pentachaeta (*Pentachaeta lyonii*). Lyon's pentachaeta formerly occurred on Catalina Island and the Palos Verdes Peninsula, but has not been seen at either location since the turn of the century. This delicate yellow member of the sunflower family is increasingly threatened by encroaching urban development.

The Santa Monica Mountains Na-

tional Recreation Area was originally mandated to acquire and protect about 125,000 acres. To date, approximately 75,000 acres have been acquired by various public agencies. Though development continues to consume valuable habitat, particularly around the periphery at lower elevations, federal, state, and local agencies are managing

protected wildlands in the Santa Monica Mountains for their natural resource values and are providing excellent opportunities for Southern Californians to enjoy these natural lands. The close interface between urban areas and wildlands makes fighting wildfires extremely difficult, impeding appropriate vegetation management and ultimately resulting in habitat losses.

Tim Thomas

This pitcher sage (Salvia spathacea), TOP LEFT, is a true sage. Another group of California mints, also called pitcher sage, are in the genus Lepechinia. • The uncommon and woolly California milk-weed (Asclepias californica), ABOVE LEFT, is the larval food plant of the monarch butterfly. • The purple flowers of prickly phlox (Leptodactylon californicum), TOP RIGHT, come alive with color as dusk approaches. • Yellow members of the sunflower family (Asteraceae) are ubiquitous in California, giving rise among botanists to derisive monikers such as ADYC, Another Darn Yellow Composite. This one, RIGHT, is a tickseed (Coreopsis bigelovii). • California live oaks (Quercus agrifolia), OPPOSITE, are reflected in a pond on Nicholas Flat off Decker School Road.

DUDLEYAS IN THE SANTA MONICA MOUNTAINS

Almost anyone who has a succulent house plant will recognize the growth form of a *Dudleya*, a member of the stonecrop family (Crassulaceae). In the wild, dudleyas are often found clinging tenaciously to hostile, rocky cliffs, thus invoking one of their common names, live-forever. Over millions of years, these striking plants have evolved survival strategies for thriving in dry environments. Their growth form is low and compact, which minimizes their exposure to the sun, and they economize on water loss by reduction of their leaves, stems, and number of stomata (leaf pores).

California is home to about thirty-nine species of dudleyas. The Santa Monica Mountains have one of the richest dudleya floras in all of California with about a dozen different species. Five are found only in these mountains. Sea lettuce (*Dudleya caespitosa*) reaches the southern limit of its distribution at Point Dume on the coast, and many-stemmed dudleya (*D. multicaulis*) reaches its northern limit in the interior rain-shadow at the eastern end of the Simi Hills. Santa Monica Mountains dudleya (*D. cymosa*) is represented by three subspecies and one as yet unnamed form; of the four dudleyas in this complex, three are endemic. The west end of the mountains contains unique microhabitats with three rare dudleyas, Conejo dudleya (*D. abramsii* ssp.

parva), Verity's dudleya (*D. verityi*), and Blochman's dudleya (*D. blochmaniae* ssp. *blochmaniae*).

Dudleyas are rare in part because they have evolved in a specialized and uncommon habitat: on isolated rock outcrops. Increasing residential development, road construction and grading, and clearing of vegetation for fire breaks in the Santa Monica Mountains threaten the survival of these species. Indiscriminate plant collecting by succulent collectors and damage from rock climbers are additional threats, particularly for the rarest species.

Tim Thomas

The west end of the Santa Monica Mountains contains unique microhabitats with several rare dudleyas, including Blochman's dudleya (D. blochmaniae *ssp.* blochmaniae) (*below left*), *Verity's dudleya* (D. verityi), ABOVE, *and marcescent live-forever* (D. cymosa *ssp.* marcescens), BELOW.

One of California's ten native box thorns (Lycium californicum), LEFT, *is the backdrop here for striking rosettes of bright green dudleya (Dudleya virens), a declining species. • Despite degradation, the wide coastal terraces of the Palos Verdes Peninsula remain unique, sheltering a few species that have their only California mainland occurrence here and are found in greater abundance on the Channel Islands. These include the white-flowered shrub, Catalina crossosoma (Crossosoma californicum),* ABOVE, *and aphanisma (Aphanisma blitoides),* BELOW, *which was once thought to be extirpated in Los Angeles County but was rediscovered in 1991. • The bright green dudleya (Dudleya virens),* BELOW LEFT, *also has its only mainland occurrence in the Palos Verdes Hills at the seaward edge of the Los Angeles Basin. It is found in greater abundance on the Channel Islands.*

PALOS VERDES PENINSULA

Palos Verdes Peninsula, formerly one of the Channel Islands, has been attached to the mainland only since the end of the last ice age. Like that of the islands, the flora of this peninsula to the south of the Los Angeles Basin developed through thousands of years of isolation. Following attachment to the mainland, the flora was altered significantly by new plant colonization and competition. Establishment of large ranches and, more recently, significantly increased suburban development and recreational use have further changed the peninsula.

The peninsula's dominant plant communities are the increasingly threatened coastal sage scrub, the rare coastal bluff scrub, a few riparian areas, and a non-native annual grassland introduced by grazing and dry farming. An uncommon clay soil plant, the small-flowered morning-glory (*Convolvulus simulans*), occurs in a limited area within the grassland.

Rare plants of the coastal sage scrub community include the Catalina mariposa lily (*Calochortus catalinae*), western dichondra (*Dichondra occidentalis*), and the prostrate annual South Coast saltscale (*Atriplex pacifica*). These species will be threatened if development continues without proper conservation of the peninsula's unique features. Surrounded by the extensive development south of Los Angeles, Palos Verdes Peninsula has once again become a biological island.

Angelika Brinkmann-Busi

MOUNT PINOS

The Chumash people call the summit of Mount Pinos Li-yikshup, meaning "center of the world." Located within Los Padres National Forest, the 8,831-foot summit of Mount Pinos serves as a landmark for miles around. Most of the mountain, located south of the San Joaquin Valley in Ventura County, is visible to the west of Tejon Pass. The bedrock of Mount Pinos consists of granite and quartz, which have de-composed to form an undulating land-scape covered with coarse gravels. Winter storms bring rain, snow, and fierce cold winds.

The harsh environment on the iso-lated summit has created a subalpine plant community far removed from other mountain peaks supporting sim-ilar vegetation. White fir (*Abies con-color*), Jeffrey pine (*Pinus jeffreyi*), pla-teau gooseberry (*Ribes* spp.), white flowering currant (*R. indecorum*), and

Parish's snowberry (*Symphoricarpos parishii*) all grow on or near the summit. Winds twist and shape limber pine (*Pinus flexilis*) near the mountaintop. Several rare plants grow on Mount Pinos, including Mount Pinos onion (*Allium howellii* var. *clokeyi*) and the dark blue Mount Pinos larkspur (*Delphinium parryi* ssp. *purpureum*).

Many tourists visit Mount Pinos throughout the year, in the summer to hike and ride mountain bikes, in winter to cross-country ski and snowboard. The U.S. Department of Defense operates a communications facility near the summit. Recognizing a need to protect sensitive plant species in this area, Los Padres National Forest has designated 510 acres as the Mount Pinos Summit Botanical Area, where plants receive additional protection.

Karen Danielsen

This meadow with conifers and shrubby lupines (Lupinus *spp.*), TOP LEFT, *receives ample water as a result of Mt. Pinos' high elevation and snowpack. • Limber pine* (Pinus flexilis), LEFT, *grows twisted and gnarled above a carpet of ground-hugging perennials that include Pursh's woolly-pod* (Astragalus purshii), *easily recognized by its densely hairy fruit. • Mountain lousewort* (Pedicularis semibarbata), ABOVE LEFT, *is a root parasite. Its name reflects an ancient belief that ingestion by stock promoted lice infestation. • As are so many others in the genus, flax-like monardella* (Monardella linoides *ssp.* oblonga), ABOVE, *is restricted in distribution and endemic to California.*

SAN GABRIEL MOUNTAINS

Surrounded by major faults, the highly dissected, complex, and seismically active San Gabriel Mountains cover some 700 square miles. Characterized by narrow, tortuous canyons and rugged peaks, the range has a sharp crestline averaging around 4,500 feet in elevation and rising to 10,064 feet at San Antonio Peak. The range separates the Santa

Red shank (Adenostoma sparsifolium), LEFT, *is also called ribbon bush because of its shredding bark. • The beautiful and rare lemon lily* (Lilium parryi), TOP, *grows in moist stream and meadow habitats. • The large genus of currants and gooseberries is known throughout the northern hemisphere. Many California species, including chaparral currant* (Ribes malvaceum), ABOVE, *are pollinated by hummingbirds. • Jeffrey pine* (Pinus jeffreyi), *white fir* (Abies concolor), *and lodgepole pine* (Pinus contorta *ssp.* murrayana), OPPOSITE TOP, *occur at much higher elevations in the mountains of Southern California than in the Sierra Nevada. • The San Gabriel Mountains,* RIGHT, *are an upthrust block of rock extending some sixty miles from west to east with a central width of twenty to twenty-five miles. Forming the northern boundary of the Los Angeles Basin, the rugged San Gabriels support a rich flora with approximately 1,200 species of vascular plants.*

Clara and Antelope valleys from the San Fernando and San Gabriel valleys in the Los Angeles Basin. Cajon Pass on the east divides this range on the Pacific tectonic plate from the larger, higher San Bernardino Mountains on the North American plate.

The mountains' wide range of elevations, precipitation, and microclimates results in a great variety of habitat types, which in turn host a fascinating diversity of plant communities. On the coastal side, slopes support sage scrub, riparian woodland, southern oak woodland, chaparral, and various forest types. From the highest elevations downward, slopes toward the desert side support several types of forest and woodlands, desert transitional chaparral, and sagebrush and desert scrub.

Over twenty rare plants are known from the San Gabriel Mountains, including the rare Mount Gleason paint-

brush (*Castilleja gleasonii*) found blooming in May and June at higher elevations on granitic soils in yellow pine forests, the rare lemon lily (*Lilium parryi*) of seeps and creeks, and the short-jointed beavertail cactus (*Opuntia basilaris* var. *brachyclada*) found along north slopes in open chaparral and in pinyon-juniper and Joshua tree woodlands.

The Angeles and San Bernardino national forests administer most of the land in the San Gabriels and are guided by policies dedicated to conserving and managing endemic species. Plants are threatened by urban development, mining, and road construction. The mountains are readily accessible to Los Angeles residents. Both national forests face serious challenges in the conservation and management of plants and animals with an ever-increasing urban populace and an ever-shrinking funding base.

Robert F. Thorne

THE PEBBLE PLAINS IN BIG BEAR VALLEY

An unusual and diverse alpine-like flora with as many as 100 plant species can be found in the desert-like environment created on the Pebble Plains, an ancient lakebed in Big Bear Valley, high in the San Bernardino Mountains. The Pebble Plains may support the highest concentration of endemic plants in all of the United States.

Big Bear Lake was part of the Mojave Desert floor when it was covered by a large Pleistocene lake. Deep clay deposits from the lake bottom persist today in pockets on the hills surrounding Baldwin Lake. The clay deposits are called the "pebble plains" because of a layer of orange quartzite pebbles pushed to the surface by freezing and thawing. Many plants of the pebble plains exhibit a low-growing, cushion-like growth form more typical of higher-elevation alpine plant life. These unusual and brightly colored species lend a rock garden aspect to this remarkable habitat. Nearly half the plant species on the plains are miniature annuals that flower and set seed in the spring, then disappear before the hot, dry summer.

Where clay deposits are crossed by springs or creeks, a unique subalpine meadow habitat is created. Here are found several of California's rarest and most threatened plants, including the endangered Big Bear checkerbloom (*Sidalcea pedata*) and slender-petaled mustard (*Thelypodium stenopetalum*).

Most of the Pebble Plains and subalpine meadows are under the protective management of the Big Bear Valley Preserve System through the efforts of the San Bernardino National Forest, The Nature Conservancy, and the California Department of Fish and Game. Management of this regional preserve, which includes some of the best remaining examples of these habitats, poses a conservation challenge because of fragile exposed plant populations. Public interest and involvement has increased with the establishment of a Friends group which has developed educational programs for the public.

Tim Krantz

Ash-gray Indian paintbrush (Castilleja cinerea), LEFT, *glows yellow, not gray, in soft evening light.* • Bird's-foot checkerbloom (Sidalcea pedata), TOP, *a state and federal endangered species, has been extirpated from much of its range by urban development, grazing, and off-road vehicles.* • Quartzite pebbles litter the thick clay on the pebble plains, ABOVE RIGHT, *a habitat found no place else in the world.* • In April and May the plains are awash with miniature plants, including the rare lilac-flowered Parish's rock cress (Arabis parishii), RIGHT, *and the widespread common yellow violet* (Viola douglasii), FAR RIGHT.

CARBONATE ENDEMICS OF THE SAN BERNARDINO MOUNTAINS

The Helendale rift zone cuts like a knife across the northeast edge of the San Bernardino Mountains, exposing fractured layers of carbonate (both limestone and dolomite) tilted at odd angles along its course, and extending along the ridges north of Big Bear Lake and Holcomb Valley. Soils developed from carbonate support a variety of plant communities, from Jeffrey pine forest at upper elevations to pinyon-juniper woodland, blackbrush scrub, and Joshua tree woodland, which appears as the steep canyons empty out onto broad alluvial fans of the Mojave Desert.

Five carbonate plants were designated by the U.S. Fish and Wildlife Service as endangered in 1994. The San Bernardino National Forest then initiated the development of a species recovery and habitat management plan that will lay the groundwork for a series of preserves. To establish permanent preserves, the voluntary relinquishment of active mining claims will need to be achieved. Efforts to reform the 1872 Mining Act to better protect endangered species and wildlife habitats will continue. Until then, establishing preserves will be difficult given the existence of many active claims.

Huge, open-pit quarries lie on the

The highly endangered Cushenbury milk-vetch (Astragalus albens), LEFT, *occurs only on San Bernardino carbonate soils. • Many plants in these carbonate communities are found on other pockets of similar soils throughout the western states, but five are endemic to the north slopes of the San Bernardino Mountains. One of these is Cushenbury buckwheat* (Eriogonum ovalifolium *var.* vineum), *with lovely silver-gray prostrate leaf clusters and flowering stems,* ABOVE AND OPPOSITE TOP, *and an intricate inflorescence. • Damage from mining is widespread,* OPPOSITE, SIDEBAR; *plants are unable to cover vertical walls left by mining operations. Vast quantities of dust from the waste piles are blown onto vegetation and roads, covering them with a quarter-inch layer of cement.*

north slope of the Big Bear Ranger District of the San Bernardino National Forest and on adjacent Bureau of Land Management lands. Existing mining operations are being expanded and new operations are being planned. Virtually all remaining carbonate is under mining claims, established in accordance with the 1872 mining laws, and the statutory rights of claim holders limit the ability of the Forest Service to regulate activities on these lands.

The Endangered Species Act does provide for certain protection through the consultation process and the development and implementation of recovery plans.

Tim Krantz and Connie Rutherford

CARBONATE LIMESTONE AND DOLOMITE SOILS

Limestone and dolomite soils are derived from calcium carbonate (calcareous) compounds deposited in the marine environment of ancient tropical coral seas. In dolomite soils magnesium is substituted in varying degrees for the calcium found in limestone, often resulting in virtually imperceptible gradations between the two. Deep deposits of limestone have been uplifted in many parts of the world, forming hard surfaces that are resistant to weathering, especially in arid climates. High-grade limestone has many uses, ranging from cement to antacids to paper whiteners and is mined extensively.

Calcareous soils are of special interest because many evolutionary relics (paleoendemics) are restricted to these soil types. *Dedeckera eurekensis*, for example, is in a genus of its own in the buckwheat family and is found only in the Eureka Valley; *Neviusia cliftonii*, in the rose family and a recent discovery growing on limestone soil in northeastern California, has its nearest relative in Alabama.

Calcicolous paleoendemics are often characterized by low seed sets when compared to the regionally dominant species. Often only a small proportion of the few seeds produced is fully viable, making these species especially vulnerable to human-mediated environmental disturbance. Because of their isolation from related species, paleoendemics possess uncommon germplasm and should be given priority for protection over neoendemics adapted to particular soils where there are close relatives in large genera.

Delbert Wiens and Edward J. King

SANTA ANA RIVER WASH

The upper Santa Ana River wash on the coastal slopes of the San Bernardino Mountains harbors one of the last remnants of a disappearing California habitat, Riversidian alluvial fan sage scrub. Here, one can hear a cactus wren's raspy call floating over the California juniper scrub and watch horned lizards scurrying across the sand. This habitat is literally armed with cholla and prickly-pear (*Opuntia* spp.), Our

plains of the coastal San Bernardino and San Gabriel mountains.

Today, the desert species of the Santa Ana River wash are disjunct from their desert of origin, separated by the high intervening San Bernardino Mountains. Over thousands of years of isolation, spineflowers, for example, have evolved to become unique species. Five species occur here, and several are extremely rare or endangered. The endangered slender-

Lord's candle (*Yucca whipplei*), and spineflower (*Chorizanthe* spp.).

The Santa Ana River wash was created by a million years of erosion and deposition of sediments as the San Bernardino Mountains uplifted along the mighty San Andreas fault, which marks the north and east boundaries of the alluvial fan. About 4,000 to 9,000 years ago, when the climate was warmer and drier than today, the Mojave Desert expanded to the west into the southern San Joaquin Valley and through the passes to the flood-

horned spineflower (*Dodecahema leptoceras*) once ranged from the San Fernando and San Gabriel valleys to the San Bernardino Valley. Today it is found from the Big Tujunga wash in Los Angeles County to the Santa Ana River wash in San Bernardino County. Several scattered pockets of this spineflower occur on the Santa Ana River in stable sandy openings, often near low, sprawling California junipers. It is easily overwhelmed by the introduced grasses that invade its habitat when disturbed or after a long interval

without flooding. Nearby, on open, recently deposited sands, the endangered Santa Ana River woollystar (*Eriastrum densifolium* ssp. *sanctorum*) grows.

The habitat of both species is being consumed by urbanization of the greater Los Angeles metropolitan area. As the washes are channelized for flood control, the remaining native vegetation is marred by huge sand and gravel quarries where the alluvial deposits of a million years are mined to make concrete and asphalt.

Tim Krantz

The desert affinities of alluvial fan sage scrub are revealed by leaf rosettes of Our Lord's candle (Yucca whipplei) and prickly-pear (Opuntia *spp.*), OPPOSITE. • *A two-foot-tall showy perennial, the endangered Santa Ana River woollystar (Eriastrum densifolium ssp. sanctorum), ABOVE, occurs on exposed, recently deposited sands in the river wash. This plant, once more common, has white-woolly foliage and sky-blue flowers over an inch long which are visited by an array of pollinators such as flies, sphinx moths, and hummingbirds. It is now known only from one extended population in the Santa Ana River watershed. • Plummer's mariposa lily (Calochortus plummerae), LEFT, has been eradicated by urban development from much of the lower-elevation Los Angeles Basin, and like much of Southern California's coastal flora, it continues to decline.*

175

SOUTH COAST AND THE PENINSULAR RANGES

The southwestern region of California rises from sea level along the coast to 10,786 feet at the top of San Jacinto Peak in the Peninsular Ranges to the east and encompasses some 7,200 square miles of Orange, Riverside, and San Diego counties. The Peninsular Ranges are formed by a broken chain of mountains, mostly granitic with occasional large outcrops of gabbro rock, and include the San Jacinto, Santa Rosa, Santa Ana, Hot Springs, Volcan, Palomar, Cuyamaca, and Laguna mountains. These ranges are older geologically and lower in elevation than the Transverse Ranges to the north.

Above 4,000 feet, this region may receive as much as sixty inches of precipitation in a year. Fog and cool breezes carry a maritime influence inland to the east-west riverine valleys of the Santa Ana, San Luis Rey, San Dieguito, San Diego, and Tijuana rivers.

Chaparral and coastal sage scrub form the dominant plant communities on plains and lower slopes. Woodlands and oak savanna occur in association with drainages at lower elevations and coniferous forests of Jeffrey, lodgepole, and limber pine become more extensive above 3,000 feet.

A unique flora results from a mixture of plants associated with Baja California and relictual elements of Sierra Nevadan origin. As various segments of the Peninsular Ranges were uplifted, plant populations shifted in response to climatic changes and were subsequently isolated, resulting in several species with widely disjunct populations. Creeping sage (*Salvia sonomensis*), for example, grows to the north in the Sierra Nevada and Coast Ranges, and again appears in the Peninsular Ranges in San Diego County.

Despite the extensive loss of biological diversity to suburban development, agriculture, and flood control, the southwestern corner of California retains a rich native flora, with some 1,525 plant species, subspecies, and varieties. Many unusual and endemic species grow here, such as the Tahquitz ivesia (*Ivesia callida*), a small, tufted, alpine-like plant in the rose family that was once thought to be extinct.

Today, the principal threat to the plant diversity of the region lies in the relentless conversion of lands for construction of homes and associated commercial activities. Vast expanses of native bunchgrass and coastal sage scrub vegetation have been

---

A field of California encelia (Encelia californica), RIGHT, *faces a coastal slope covered by soft chaparral or coastal sage scrub at Point Loma in San Diego County.* • *Chia (Salvia columbariae ssp.* columbariae), BELOW LEFT, *is common in openings in chaparral below 3,600 feet.* • *False lupine (Thermopsis macrophylla),* BELOW RIGHT, *is found further to the north in the Sierra Nevada but is not present in the Transverse Ranges and reappears in the Cuyamaca Mountains of the Peninsular Ranges.* • *Early spring brings fields of California poppies and goldfields to the Lake Henshaw area in San Diego County,* ABOVE RIGHT.

eliminated. Agricultural clearing has been devastating, especially since the invention of drip irrigation, which has allowed the planting of crops on otherwise non-arable slopes. Flood control projects have destroyed riparian habitats by damming stream flows and eliminating dynamic flooding. Past logging practices have diminished the region's forests. New regional planning efforts are attempting to identify valuable natural lands and mechanisms to conserve them for future generations.

R. Mitchel Beauchamp

UPPER NEWPORT BAY SALT MARSH

Today, salt marshes fringe part of Upper Newport Bay, a bay estuary fed by fresh water from the San Diego Creek watershed in Orange County, smaller tributaries, and urban runoff. Only about 700 acres of fully tidal salt marsh remain in Upper Newport Bay, about five percent of a once vast wetland system.

Salt marsh vegetation is distributed in subtle elevational tiers in response to tidal flows and salinities. At lower, more frequently flooded elevations, there are large stands of tall cordgrass (*Spartina foliosa*) at the edge of the mudflat. Next are found broad stands of the common perennial pickleweed (*Salicornia virginica*), the commonest plant in California's tidal marshes. At mid-elevation in this southern marsh, the rare salt marsh bird's-beak (*Cordy-*

lanthus maritimus ssp. *maritimus*) occurs with annual pickleweed (*Salicornia bigelovii*) and common saltwort (*Batis maritima*). In places, both pickleweed species are covered by the orange threads of a parasitic dodder (*Cuscuta salina*).

Above the salt marshes, pampas grass and other aggressive non-native plants have invaded most of the bluff areas. Predation by the introduced red fox threatens native occupants of the salt marsh, especially the endangered light-footed clapper rail and the California least tern. Upland development results in extensive siltation and eventual wetland loss and massive invasions of weedy species in upland areas. Protection of the wetlands depends directly on proper management of urban runoff and siltation.

The California Department of Fish and Game manages the Upper Newport Bay Ecological Reserve (752 acres) and Orange County manages the Upper Newport Bay Regional Park (138 acres). The bay estuary lies immediately downstream from the University of California Natural Reserve System's San Joaquin Freshwater Marsh Preserve, resulting in over 1,090 acres of natural wetland habitat conserved in the midst of one of California's most densely populated regions.

Peter Bowler

The endangered salt marsh bird's-beak (Cordylanthus maritimus *ssp.* maritimus), ABOVE, *brightens the marsh with its whitish flowers in early summer. The ability to excrete salt (seen as crystals) and dense hairs that reduce water loss enable this plant to live in its harsh environment.* • *A large stand of the salt marsh bird's-beak,* RIGHT, *grows along the marsh margin near a bayside perimeter road along with other upper-elevation plants, including alkali-heath* (Frankenia salina), *salt cedar* (Monanthochloe littoralis), *saltgrass* (Distichlis spicata), *and glasswort* (Salicornia subterminalis).

TORREY PINE FOREST

Found only on the summer fog-shrouded coastal bluffs of San Diego County near La Jolla and Del Mar and one small area of Santa Rosa Island, Torrey pine (*Pinus torreyana*) is North America's rarest pine. It grows on unusual sandstone formations along with a diverse chaparral community that includes the endemic Del Mar manzanita (*Arctostaphylos glandulosa* ssp. *crassifolia*) and short-leaved dudleya (*Dudleya blochmaniae* ssp. *brevifolia*). Like its nearest relatives, bigcone pine (*Pinus coulteri*) and gray pine (*P. sabiniana*), Torrey pine has enormous cones weighing a pound or more and armed with stout scales. Growing to only fifteen or twenty feet in height, Torrey pines slowly develop into strangely sculptured asymmetric forms from the effects of salt spray and prevailing sea winds. In this harsh environment, they are relatively short-lived.

Bark beetle infestations and urban-induced fires pose a grave danger to the limited remaining pine forest. In the early 1900s Ellen Scripps of newspaper fame created one of the nation's earliest state reserves to protect the Torrey pine. Today all the remaining trees on the mainland are either in the state reserve or within the City of San Diego's open space holding at Crest Canyon. Long-term protection will depend on the skill of park managers.

R. Mitchel Beauchamp and Craig H. Reiser

Torrey pine (Pinus torreyana), ABOVE RIGHT, *is a relict of a once larger, more widespread population that has become restricted by climatic changes and more recently by urbanization.* • *Cones of the Torrey pine,* BELOW, *are massive, sometimes weighing over a pound.* • *Torrey pines typically grow on sandstone formations,* RIGHT, *with an understory that includes the endemic coast white-lilac* (Ceanothus verrucosus) *and Cleveland sage* (Salvia clevelandii).

SAN DIEGO VERNAL POOLS

Small, seasonal vernal pools were once a common sight on the coastal mesas of southern San Diego County. Today, in large part because the remnant alluvial or marine terraces found on the flat mesa tops are prime areas for urban development, more than ninety-five percent of the pools have been lost. Pools can remain filled for three to five months, although some pools may not fill at all in dry years. San Diego's vernal pools range from fifty to 2,500 square feet in size and from a few inches to a foot or more deep.

Over one-half of the plant species most characteristic of San Diego's increasingly rare vernal pools are endemic to them and another quarter are restricted to them. Pool species have specialized adaptations enabling them to tolerate lengthy periods of inundation. These plants, however, are not truly aquatic, so they cannot complete their life cycle where water stands for more than six months. Grassland plants, particularly exotics, are very dense immediately above the elevation of highest water level in the pools, while native vernal pool species are infrequent above the high-water level. Vernal pool species thus occupy an intermediate habitat, neither terrestrial nor aquatic, which may have led to a proliferation of rare species.

Urban development and agriculture have severely depleted the vernal pool habitat of San Diego County. Many of the remaining pools have been extensively damaged by off-road vehicle traffic, grazing, dumping, and altered drainage patterns. When im-

Toothed downingia (Downingia cuspidata), OPPOSITE, BOTTOM LEFT, grows in vernal pools in Miramar in San Diego County. Because vernal pools in southern San Diego County are located on flat mesa tops, prime areas for urban development, more than ninety-five percent of pools have been lost. • Vernal pool species have special adaptations to live under variable moisture conditions. Early in the season, standing water triggers growth of unbranched elongate stems and simple undissected leaves, often soft and hollow. Later, in drier conditions, the plants become fibrous, stems branch, and emerging flower clusters have hairy or prickly parts. In dry years, plants develop directly into a terrestrial late-season form. Two prominent restricted species include the rare San Diego button-celery (Eryngium aristulatum var. parishii), OPPOSITE TOP, a member of the carrot family and sometimes called coyote thistle, and the endangered annual San Diego mesa mint (Pogogyne abramsii), a small fragrant mint with rose-colored flowers shown here, LEFT, growing with the larger Orcutt's brodiaea (Brodiaea orcuttii). • Orcutt's brodiaea, OPPOSITE, RIGHT, is a rare perennial, infrequently found in various grassy habitats.

pervious claypans are pierced, pools are destroyed. Preservation efforts in San Diego County have been largely unsuccessful and most of San Diego's remaining pools, some badly degraded, are on federal lands where they are subject to federal endangered species and wetlands laws. Conservation ef- forts now are focused on protecting the remaining privately owned pools and restoring those on federal lands.

Ellen T. Bauder

SAN DIEGO THORNMINT

San Diego thornmint (*Acanthomintha ilicifolia*) is a member of a small group of annual mints limited to patchily distributed clay soils in San Diego County and northwestern Baja California, Mexico. These special soils are of sedimentary origin or are derived from gabbro rocks. The thornmint can be found growing with two other rare plants, Parry's tetracoccus (*Tetracoccus dioicus*) and Dehesa beargrass (*Nolina interrata*), both of which are also restricted to gabbroic soils.

Thornmints appear to be particularly sensitive to yearly fluctuations in rainfall. In some years they are abundant on patches of suitable soils, and in other years few plants reach flowering stage. One of the reasons for their limited distribution may be an inability to compete with other plants off the clay soils.

San Diego thornmint is difficult to protect because its natural habitat is fragmented and may be overlooked in large-scale conservation planning efforts. Fewer than half the historically known populations exist today. Only a dozen of the thirty-seven remaining populations are on protected lands. Cleveland National Forest botanists fenced a population on Viejas Mountain using materials provided by the California Department of Fish and Game to protect it from damage by stray livestock.

Most of the San Diego thornmint populations are on private property, much of which is slated for development. Urbanization, weed infestation, livestock grazing, and off-road vehicles all pose threats to the future of this rare species.

Ellen T. Bauder

San Diego thornmint germinates with the onset of winter rains and grows to less than a foot in height. Its presence is often detected long before it is seen because its toothed leaves are quite aromatic. In late spring, white flowers, tinged with rose or lavender, are produced in head-like clusters at leaf nodes. Flower bracts and sepals are spiny, hence its scientific name, Acanthomintha ilicifolia, which means thornmint.

COASTAL SAGE SCRUB

A unique community of plants, extending from San Francisco into Baja California, has evolved in response to Mediterranean-like summer fog and winter rains along the coast. Coastal sage scrub, sometimes called "soft chaparral," is found from sea level to 3,000 feet in more southerly inland portions of its range, such as the coastal plains and foothills of the Transverse and Peninsular ranges, the Sierra San Pedro de Martir in Baja California, and on the off-shore islands of California and Baja California.

In contrast to the inland drier evergreen "hard chaparral" community with which it intergrades, the leaves of many members of the coastal sage scrub community drop during summer drought and are replaced by smaller and fewer leaves. The soft chaparral root system is shallower, and its canopy more open, allowing for a greater number of associated grasses, annuals, and succulents than found in hard chaparral areas.

Like chaparral, coastal sage scrub is fire prone and has evolved to withstand periodic burning. During the first post-burn years, an herbaceous cover of specialized fire-following annuals appears, gradually replaced by a woody shrub cover. Too frequent fires can lead to a conversion from coastal sage scrub to a grassland community.

Early European settlement markedly reduced coastal sage scrub habitat, which is particularly vulnerable to agricultural development since it is on relatively fertile lowlands. During the

NATURAL COMMUNITY CONSERVATION PLANNING

Through its unique Natural Community Conservation Planning (NCCP) Program, the State of California has developed a process for addressing threats to natural habitats and the species that inhabit them. In partnership with the U.S. Fish and Wildlife Service and local jurisdictions, the California Department of Fish and Game has developed a community level planning process to assure the continued existence of valued biological resources. The coastal sage scrub community was the first selected for this NCCP program.

In seeking to avoid typical "development versus environment" conflicts, local authorities, developers, landowners, conservation organizations, and scientists work together to formulate ways to conserve habitat while allowing appropriate development. The goal is to identify and develop large interconnected reserve systems within a planning area early on, rather than relying upon individual plans for development projects.

The State selected the threat-ened coastal sage scrub plant community as a pilot for the NCCP program because of its substantial historic losses, its high continuing development pressures, and the large number of threatened and endangered plants and animals the community harbors. Well known sensitive animals of the coastal sage scrub are the California gnatcatcher, San Diego cactus wren, and orange-throated whiptail lizard.

The planning area for the coastal sage scrub NCCP program is approximately 6,000 square miles, and includes portions of Los Angeles, San Bernardino, Riverside, Orange, and San Diego counties. Several multi-species reserves have been approved to date, including a 37,000-acre reserve system that specifically protects thirty-nine sensitive species in a 209,000-acre planning area in Orange County. By 1997 reserves in southwestern San Diego County covering 172,000 acres in a 582,000-acre planning area protecting eighty-five sensitive species are anticipated.

Bill Tippets

past two decades rapid urbanization has increasingly displaced coastal sage scrub. Estimates of historic losses in San Diego, Orange, and Riverside counties range from sixty-six to ninety percent. In addition, many of the remnants of coastal sage scrub communities are degraded by grazing, nonnative weed invasion, frequent fires, recreational activities, military training exercises, and possibly air pollution. Nearly 100 plants and animals of the coastal sage scrub community are classified as rare, sensitive, threatened, or endangered. Both state and local efforts to conserve large blocks of this once widespread habitat are underway.

John F. O'Leary

In the drier more southerly portion of coastal sage scrub distribution, cactus species (Opuntia sp.), OPPOSITE TOP, become more common. • Three species of sage: white (Salvia apiana), shown here growing with Indian paintbrush (Castilleja sp.), LEFT, black (S. mellifera), and purple (S. leucophylla) all grow in the coastal sage community along with California sagebrush (Artemisia californica). • The flowers of California encelia (Encelia californica) and San Diego County viguiera (Viguiera laciniata) add colorful yellows to the landscape, ABOVE, amid the cinnamon brown shades of the buckwheats (Eriogonum fasciculatum, E. cinereum), RIGHT.

WESTERN RIVERSIDE COUNTY

A surprisingly colorful and varied flora of alkali scrub, grassland, and vernal pool communities thrives in strongly alkaline soils of lowland floodplains of the San Jacinto River and Hemet Valley in western Riverside County. Early in spring, vernal pools fill with mouse-tail (*Myosurus minimus*) and woolly-heads (*Psilocarphus brevissimus*). As the pools begin to dry and the mouse-tail fades, an explosive display of goldfields (*Lasthenia californica* and *L. glabrata* ssp. *coulteri*) occurs, along with a carpet of moisture-loving alkali grassland species.

Three rare saltbushes thrive in the alkali scrub communities: San Jacinto Valley crownscale (*Atriplex coronata* var. *notatior*), Parish's brittlescale (*A. parishii*), which grows nowhere else, and one of the last mainland populations of South Coast saltscale (*A. pacifica*). By mid-May the grasses have turned brown and all that remains of the floral show is the gray-green saltbush, the dying stems of tarplant, and patches of reddish sea-blite (*Suaeda moquinii*). As these alkali communities await the return of winter rains, they look deceptively uninteresting.

Clay soils, especially along the western edge of the county in the Elsinore Mountains, Temescal Valley, and Gavi-

Ground pink (Linanthus dianthiflorus), ABOVE, *is a common sight in early spring on open slopes below 3,000 feet in western Riverside County.* • *The threatened Munz's onion* (Allium munzii) *(below left) is endemic to clay soils of western Riverside County.* • *Several seriously threatened plants, such as the lovely thread-leaved brodiaea* (Brodiaea filifolia), BOTTOM LEFT, *occur in vernal pools and associated grasslands.* • *A view of Mystic Lake,* BELOW, *can be seen from the badlands in western Riverside County.*

lan Hills, also provide significant habitat for rare plants. These areas are typically dominated by native grasslands, open coastal sage scrub, or mixed grassland-covered sage scrub communities.

Agriculture and urbanization have destroyed all but 7,000 of the original 30,000 to 40,000 acres of alkali habitat in western Riverside County. Remaining habitat is disked for fire control or weed abatement, disrupting drainage patterns and increasing weedy species. Proposals to channelize the San Jacinto River would end natural flooding. Other proposals plan urban development over much of the remaining alkali communities. Only a small fraction of these communities are protected within the California Department of Fish and Game's San Jacinto Wildlife Preserve.

Fred M. Roberts, Jr.

VAIL LAKE

Vail Lake, a man-made reservoir, is located in the foothills east of Temecula at the northern base of the Palomar Mountains in western Riverside County. It was built in the 1930s in a natural basin at the confluence of Temecula Creek and its major tributaries.

Vail Lake ceanothus (*Ceanothus ophiochilus*) was unknown to botanists until 1989. Its discovery on a twenty-acre outcrop of unique gabbroic soils on Vail Mountain occurred in early spring of 1989 when its pale blue to pinkish flowers distinguished this shrub from surrounding chaparral vegetation. Another population has since been found approximately three miles to the south on thirty acres of the Agua Tibia Wilderness, also on gabbroic soils. Here it is protected from most human disturbances, but it hybridizes readily with the widespread hoaryleaf ceanothus (*C. crassifolius*). The Vail Lake population grows entirely on private land and remains vulnerable to future development.

Willowy Nevin's barberry (*Mahonia nevinii*) has been known from the Vail Lake region since the 1930s, before the natural basin was dammed. This handsome shrub, now used in drought tolerant landscaping, is restricted to Vail Lake and a handful of small natural occurrences in northwestern Los Angeles and San Bernardino counties. The largest populations occur on privately owned land at the southwestern end of Vail Lake. Throughout its natural range, Nevin's barberry is generally found as mature individuals. Seedlings are rarely found, though they are not uncommon in horticultural settings. Scattered plants are found in the Cleveland National Forest and on Bureau of Land Management land on the summit of Oak Mountain; the remainder grow on private lands and are unprotected.

A third floral gem, the slender-horned spineflower (*Dodecahema leptoceras*) is a small, delicate annual. It is found in open alluvial scrub habitat, once common throughout the floodplains of the Santa Clara, San Gabriel, Santa Ana, and Santa Margarita river systems. This increasingly rare plant grows on alluvial bench habitats that are periodically scoured with natural flood waters or on gravelly openings in chaparral. Because of river channelization, sand and gravel mining, and urbanization, it is now limited to a few small, fragmented populations. An estimated sixty percent of all known plants of this species occur in the Vail Lake region on privately held lands. Only a few populations are protected on lands administered by the Cleveland National Forest.

Steve Boyd

---

*Two of California's rarest botanical treasures near Vail Lake in southwestern Riverside County: Nevin's barberry (*Berberis nevinii*), BELOW, and Vail Lake ceanothus (*Ceanothus ophiochilus*), BOTTOM. The uncanny resemblance of the leaves of this ceanothus to those of the ubiquitous chamise (*Adenostoma fasciculatum*) is a fine example of ecological convergence—they live in similar habitats—and may have delayed discovery of this Riverside County endemic.*

SANTA ROSA PLATEAU

The Santa Rosa Plateau delights a visitor's eye with its rolling grasslands, mesas, oak woodlands, chaparral, and coastal sage scrub-covered hills. This 45,000-acre plateau is located at the southeastern end of the Santa Ana Mountains in the Peninsular Ranges of Riverside County.

Except for vernal pool plants, the flora of the Santa Rosa Plateau is similar to that of the rest of the Santa Ana Mountains and other nearby southern mountains. The plateau is home to one of the best remaining stands of the rare and handsome Engelmann oak (*Quercus engelmannii*), which dominates portions of its mesas and savannas.

Rare plants found in the vernal pools include the spiny San Diego button-celery (*Eryngium aristulatum* var. *parishii*), California Orcutt grass (*Orcuttia californica*), Parish's meadowfoam (*Limnanthes gracilis* ssp. *parishii*), thread-leaved brodiaea *(Brodiaea filifolia)*, and the purple to blue-flowered Orcutt's brodiaea (*B. orcuttii*).

With urban expansion of Riverside and Los Angeles into the area, and because of the relatively unspoiled beauty of the grasslands, oak woodlands, vernal pools, and streamside woods, The Nature Conservancy

(TNC) purchased 3,100 acres in 1984 to create the Santa Rosa Plateau Preserve. The preserve has since been expanded to about 7,000 acres, the result of a landmark cooperative endeavor involving the Metropolitan Water District of Southern California,

the Riverside County Regional Park and Open-Space District, the Department of Fish and Game, and TNC. It is currently being managed by TNC to protect the natural communities and rare flora and to educate the public.

Earl Lathrop

Englemann oak (Quercus engelmannii), ABOVE, *dots the savannas of the Santa Rosa Plateau, where there has been a long history of use by Native Americans and for cattle grazing. • Heavy, expansive clay soils on the mesa tops trap winter rains and form vernal pools that turn blue in spring when the beautiful downingia* (Downingia bella), OPPOSITE, ABOVE LEFT, *blooms. • Golden-rayed pentachaeta* (Pentachaeta aurea), OPPOSITE, BOTTOM LEFT, *is frequent in open areas of the 2,000-foot plateau, which is bordered on all sides by steep, chaparral-clad hills. Many of these hillsides outside the reserve have recently been converted to avocado orchards.*

SAN DIEGO COUNTY'S PENINSULAR RANGES

The Peninsular Ranges in San Diego County include the Palomar, Cuyamaca, and Laguna mountains. These ranges are generally lower in elevation, less steep, and, because of winter storm patterns, moister than Peninsular and Transverse ranges to the north. They are distinguished by their many valleys and large meadows where a surprising number of rare and endemic plants can be found. Cuyamaca larkspur (*Delphinium hesperium* ssp. *cuyamacae*) occurs on Palomar and Cuyamaca mountains in the grassy meadows, along with San Diego gumplant (*Grindelia hirsutula* var. *hallii*), endemic to San Diego County, and San Bernardino bluegrass (*Poa atropurpurea*).

The upper, windward slope of Palomar Mountain is one of the wettest locations in Southern California. Its vegetation is a heavy growth of mixed coniferous forest with white fir, big-cone Douglas-fir, incense cedar, and California black oak. The scarlet-flowered Hall's monardella (*Monardella macrantha* ssp. *hallii*) grows in the shade of the forest. In the openings one finds the rare, pink Orcutt's linanthus (*Linanthus orcuttii*), purple Palomar monkeyflower (*Mimulus diffusus*), and yellow Cleveland's bush monkeyflower (*Mimulus clevelandii*). In spring, lemon lily (*Lilium parryi*) adorns wet meadows with handsome yellow blooms along with golden violet (*Viola aurea*); both are rare.

Cuyamaca Peak consists largely of gabbro rock. In addition to 500-year-old sugar pine (*Pinus lambertiana*), it supports Cuyamaca cypress (*Cupressus arizonica* var. *stephensonii*), Orcutt's bro-

The fragrant Cleveland sage (Salvia clevelandii), OPPOSITE TOP, *is found in the chaparral stands of McGinty Mountain.* • *The rare and endemic Laguna Mountains aster* (Machaeranthera asteroides *var.* lagunensis), OPPOSITE BOTTOM, *blooms on the edges of meadows on Laguna Mountain in the shade of Jeffrey pine* (Pinus jeffreyi) *after summer rains.* • *From McGinty Mountain,* LEFT, *one can look over the Peninsular Ranges, shown here emerging from coastal fog.* • *Desert phlox* (Phlox austromontana), BELOW, *grows in dry, rocky areas in pinyon-juniper forests.* • *Goldfields* (Lasthenia spp.) *and cream cups* (Platystemon californicus), BOTTOM, *grace open meadows in the Laguna Mountains.*

diaea (*Brodiaea orcuttii*), Cuyamaca raspberry (*Rubus glaucifolius* var. *ganderi*), and San Diego County alumroot, (*Heuchera rubescens* var. *versicolor*).

The Laguna Mountains are drier than the nearby Cuyamaca and Palomar mountains. The reddish Mount Laguna alumroot (*Heuchera brevistaminea*) grows on the rocky cliff-like eastern scarp along with the rare Laguna Mountains goldenbush (*Ericameria cuneata* var. *macrocephala*), which

also occurs on the top of Hot Springs Mountain.

Much of the Palomar and Laguna mountains is managed by the Cleveland National Forest; Cuyamaca and several adjacent peaks are in the Cuyamaca Rancho State Park. However, important habitat areas such as the large meadows around Cuyamaca Lake and the valleys on Palomar Mountain remain in private ownership.

Thomas Oberbauer

McGinty Mountain

Among the coastal foothills of the Peninsular Ranges in southwestern San Diego County are several low but botanically interesting peaks. These include McGinty, Dehesa, Viejas, and Poser mountains and Sycuan Peak. All are formed by outcrops of gabbroic rock. The toxic soils of these gabbroic islands restrict vegetation to gabbroic-tolerant species that make up a spe-cialized natural community, found on-ly here, called mafic southern mixed chaparral.

Two of the best examples of this kind of chaparral are found on Mc-Ginty Mountain, a large, spreading 2,183-foot mountain with steep, dark red ridges, and Sycuan Peak, a promi-nent reddish, conical 2,801-foot peak; both are located between the rural community of Jamul and the Sweet-

water River Valley. Here thrives a unique assemblage composed of chamise (*Adenostoma fasciculatum*), the fragrant and handsome Cleveland sage (*Salvia clevelandii*), mission manzanita (*Xylococcus bicolor*), Our Lord's candle (*Yucca whipplei*), laurel sumac (*Malosma laurina*), and the rare, gabbro-loving *Tetracoccus dioicus*.

Much of McGinty Mountain is threatened by residential development and associated utilities. Since 1988 new homes have been built up to the ridgeline and on the southern base of the mountain. Fortunately, some portions of the uppermost ridges and peak of McGinty Mountain have been acquired and are managed as a collective ecological reserve by The Nature Conservancy, the California Department of Fish and Game, The Environmental Trust and the County of San Diego.

Prior to 1995, over eighty percent of Sycuan Peak was in private ownership with plans for residential development. Since that time, the California Department of Fish and Game has acquired some sixty percent of this area as an ecological reserve, including the actual peak—habitat for Dehesa nolina (*Nolina interrata*), Gander's butterweed (*Senecio ganderi*) and other rare plants. The lowermost northern slopes of Sycuan Peak were acquired by The Nature Conservancy in 1991, along with a portion of the Sweetwater River Valley. This area was later sold to the Department and is today the Sweetwater River Ecological Reserve.

James C. Dice, Harold A. Wier, and John W. Brown

The peak of McGinty Mountain, TOP LEFT, *is home to one of the two largest populations of the rare Dehesa nolina* (Nolina interrata), *which is restricted to gabbroic soils in southwestern San Diego County and in a small adjacent area in northwestern Baja California.* • *The rare gabbro-loving* Tetracoccus dioicus, FAR LEFT, *is related to the giant candelabra euphorbias of East Africa.* • *Dehesa nolina,* LEFT, *flowers profusely in years following wildfires, producing three- to four-foot-tall stalks of small, cream-colored flowers. Plants are either male or female and the presence of both is required for viable seed to be produced.* • *McGinty Mountain is also home to the rare Gander's butterweed* (Senecio ganderi), TOP RIGHT, *a handsome woody-stemmed perennial that produces rosettes of purple-tinged green leaves in late winter and attractive heads of yellow daisy flowers in late April or early May.* • *Weed's mariposa lily* (Calochortus weedii), ABOVE, *is common on the open slopes of McGinty Mountain.*

191

CUYAMACA VALLEY

The Cuyamaca Valley, a montane valley on the eastern cusp of the Peninsular Ranges in southern San Diego County, provides a special and limited habitat for plants. In 1886, the outlet from this valley was dammed by the San Diego Flume Company to create a reservoir, and altered in the 1960s to turn it into an artificial fishing lake as well. The reservoir eliminated many of the valley's meadows and temporary wetlands, including a seasonal or vernal lake known as *la laguna que se seca*. Cuyamaca Valley's slopes are covered by mixed coniferous and deciduous forests. Chaparral and sage scrub cover foothills that descend westward to the Pacific Ocean, while to the east a steep escarpment plunges to the Colorado Desert at Anza-Borrego State Park.

Despite more than a century of mining, logging, and grazing, some of the meadows probably look today much as they did 150 years ago. Drainage from a surrounding bowl of hills into meadows and temporary wetlands supports a wide array of plants, including three rare species, all adapted to life in temporary wetlands: Cuyamaca Lake downingia (*Downingia concolor* var. *brevior*), Parish's meadowfoam (*Limnanthes gracilis* ssp. *parishii*), and Cuyamaca larkspur (*Delphinium hesperium* ssp. *cuyamacae*).

Downingias and meadowfoams are small but beautiful wildflowers associated with vernal pools and lakes from San Diego County to the Central Valley and Coast Ranges. Interestingly, the closest relatives of two species growing at Cuyamaca Lake are found from 500 and 800 miles northwest. By contrast, the larkspur, a perennial, is found at slightly higher elevations and produces a thickened tap root that helps it survive hot, dry summers. It produces a tall, handsome raceme of

Cuyamaca Lake downingia (Downingia concolor *var.* brevior), BELOW, *is a small soft-stemmed annual that germinates only when inundated with water and temperatures are cool. Its conical stems, like wagon wheels in cross section, allow air to diffuse to the roots so the plants can "breathe" while they are under water. In the spring, downingias produce masses of dark blue flowers, sometimes giving the appearance of standing water long after surface water is gone.* • *In May, Parish's meadowfoam* (Limnanthes gracilis *ssp.* parishii), RIGHT, *covers the meadow like froth and often mingles with downingia in moist areas and grassy openings.* • *Oaks predominate at the edge of a wet meadow at Cuyamaca State Park,* TOP RIGHT. • *Spring brings bright washes of tidy-tips* (Layia platyglossa), FAR RIGHT, UPPER MIDDLE, *interspersed with checkerbloom* (Sidalcea malvaeflora *ssp.* sparsifolia) *in meadows.* • *Another favorite floral display contains white evening primrose* (Oenothera *spp.*) *and grape soda lupine* (Lupinus excubitus), FAR RIGHT, LOWER MIDDLE. • *Checkerblooms,* FAR RIGHT, BOTTOM, *are highly variable and identification is often difficult. This checkerbloom is* Sidalcea malvaeflora *ssp.* sparsifolia.

blue to purple flowers from late June through July.

Over half of Cuyamaca Valley is protected as part of the California State Park System and by the Helix Water District, which leases a large part of its holdings to the Cuyamaca Lake Recreation and Park District. Grazing, construction, and recreational activities are threats to the showy spring wetlands and meadows. The California Departments of Fish and Game and Parks and Recreation, U.S. Fish and Wildlife Service, and the U.S. Forest Service are working on an agreement that will resolve conflicts among varied land uses.

Ellen T. Bauder

OTAY MOUNTAIN AND METAVOLCANIC PEAKS

The foothill and low-elevation peaks in the Peninsular Ranges of San Diego County, Santa Ana, San Miguel, Jamul, and Otay mountains are remnants of a chain of ancient volcanoes from which metavolcanic soils form. These climatic islands receive additional moisture from fog because the peaks rise above lower layers of air inversion, resulting in a rich and diverse chaparral community.

Several endemic plants occur on the chaparral slopes of these peaks, typically dominated by chamise (*Adenostoma fasciculatum*). Of particular interest is Cleveland's monkeyflower (*Mimulus clevelandii*), a tall, yellow-flowered herbaceous monkeyflower that represents a "missing link" between herbaceous monkeyflowers placed in the genus *Mimulus* and shrubby monkeyflowers sometimes called *Diplacus* by taxonomists. Other interesting members of this diverse chaparral community include Mexican flannelbush (*Fremontodendron mexicanum*), Otay Mountain lotus (*Lotus crassifolius* var. *otayensis*), southern mountain misery (*Chamaebatia australis*), and San Miguel savory (*Satureja chandleri*). Rare stands of Tecate cypress (*Cupressus forbesii*) intergrade in-

to the chaparral community on part of Otay Mountain.

Otay Mountain is currently managed by the U.S. Bureau of Land Management; however, under wilderness study status, it will be transferred to the U.S. Forest Service.

R. Mitchel Beauchamp

Red shank or ribbon bush (Adenostoma sparsifolium), BOTTOM LEFT, *is widespread in San Diego chaparral, and occurs on Otay Mountain with its relative, chamise* (A. fasciculatum), *a major component of the chaparral community. Here, it is growing in nearby Cameron Valley.* • *Cleveland's monkeyflower* (Mimulus clevelandii), ABOVE, *has been used frequently in the horticultural trade as parent stock for creating handsome monkeyflower hybrids.* • *The extremely endangered small-leaved rose* (Rosa minutifolia), BELOW, *reaches its northern limits here and is known in California from only one population on the Otay Mesa.* • *Named for John C. Fremont, Mexican flannelbush* (Fremontodendron mexicanum), ABOVE RIGHT, *adds bright spots of yellow to chaparral slopes in spring.* • *Twining bursts of spring color from the San Diego sweetpea* (Lathyrus vestitus ssp. alefeldii), RIGHT, *are common in San Diego chaparral.*

TECATE CYPRESS

A relic from the time when glaciers covered the continent, Tecate cypress (*Cupressus forbesii*) survives in four stands on north-facing mountain slopes and canyons in Southern California and in a few small stands in Baja California, Mexico. Each stand is unique.

In San Diego County the largest Tecate cypress forest, covering 6,000 acres and managed by the U.S. Bureau of Land Management, is found on Otay Mountain. The tallest known tree, sixty feet tall, is in Guatay Mountain's sixty-four-acre stand. The most endangered cypresses grow on Tecate Peak, where more than half of the known stand has been lost in frequent, human-caused fires. In Orange County, the oldest Tecate cypress, 250 years old, clings to the slopes of Fremont Canyon. The largest tree, with a height of forty-five feet, has a thirty-five foot diameter dripline and a trunk circumference of eight feet. It thrives in the deep, rich soil of Coal Canyon, and like other Orange County Tecate cypresses, it is genetically distinct from trees of the southern stands in San Diego County.

Tecate cypress depends on intermittent fire for reproduction. The resin seal on the cone provides an insulating charcoal layer for cones and their seeds during a fire that kills the trees. Afterwards, cones gradually open and their seeds drop to the ground, which is open to sunlight. However, if fires occur more often than every thirty-five to sixty-five years, cones cannot produce sufficient seed for the stand to be maintained.

Remaining stands of Tecate cypress are not fully protected. The Tecate Peak stand is in the most danger because of too frequent fires. Portions of stands at Otay Mountain and in Orange County are still privately owned and are threatened by encroaching development.

Connie Spenger

With bright green leaves and reddish bark, Tecate cypress trees, BELOW, vary from Christmas-tree shaped in youth to rounded or picturesquely contorted in age. A member of an ancient family of plants, Tecate cypress once grew from Orange County half-way down the peninsula of Baja California, Mexico. As the continent warmed 10,000 years ago, Tecate cypress remained in small, cooler sites where there was abundant soil moisture, sometimes leaving its seeds, cones, ABOVE, and bark in places such as the La Brea tar pits. In the floristic islands where it now grows, an assemblage of plants such as rare pitcher sages (Lepechinia spp.) and milk-vetches (Astragalus spp.) provides windows into the plant life of past geologic ages.

DESERTS

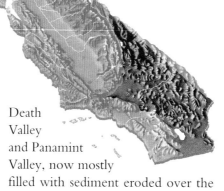

ew deserts in the world match the physical and biological splendor of those of California. Deserts cover nearly one-quarter of the state, including the western fringe of the Great Basin, most of the Mojave, and the northwestern portion of the Sonoran Desert—the Colorado Desert. Differing combinations of climate, elevation, soils, and topography have given each of California's desert landscapes—the Great Basin, the Mojave, and the Colorado—its own rich specialized flora and vegetation. About 2,000 native plant species occur in these deserts, representing about seventy percent of the state's native plant families.

East of the Sierra Nevada, the Great Basin is a high, cold desert. It is landlocked, with all its drainages ending in alkaline lakes and playas. Largely dominated by sagebrush and pinyon-juniper vegetation, the Great Basin is composed of north-south trending fault-block mountains such as the White and Warner ranges. These mountains are separated by deep troughs such as

Death Valley and Panamint Valley, now mostly filled with sediment eroded over the millennia from the surrounding desert ranges. As recently as 4,000 years ago, these troughs were filled with large, deep lakes. To the south is the Mojave Desert, a hot desert dominated largely by creosote bush scrub. The distinctive Joshua tree is a conspicuous indicator of the Mojave's mid-elevations. On its southwestern edge, the Mojave Desert is separated from the lower and even hotter Colorado Desert by the San Bernardino and Little San Bernardino Transverse ranges.

The 56,000 square miles of desert east of the Sierra Nevada and Peninsular ranges are composed of such varied land forms as rugged mountains, rounded domes, and table-topped mesas; mud hills and badlands; pumice fields, lava flows, cinder cones, and hot springs; large, saline, dry lakes usually called playas, and an inland saline sea well below sea level, the Salton Sea. These diverse landforms result in some of the steepest gradients in North America from White Mountain Peak (14,242 feet) to Badwater (-282 feet), less than 100 miles apart.

There are also rivers, though most, such as the Mojave, are intermittent and disappear into the desert sands— into great dune systems such as the Eureka, Kelso, and Algodones, and smaller crescentic dunes west of the Salton Sea. Only the Colorado River flows continuously. Active faults, such as the powerful San Andreas, produce many earthquakes each year.

Recent years have seen a drastic increase in human activity in the desert's fragile ecosystems. Water diversion,

California barrel cactus (Ferocactus cylindraceus), BOTTOM LEFT, *is a conspicuous feature of both the Colorado and the Mojave deserts. • In desert mountain ranges, separated by undrained basins and washes, varnished desert pavement, and alluvial gravel fans, barrel cactus, silver cholla* (Opuntia echinocarpa), *Spanish bayonet* (Yucca baccata ssp. baccata), *and a few feathery creosote bushes* (Larrea tridentata) *frame the Providence Mountains,* RIGHT. *The genus name,* Opuntia, *for this beavertail prickly pear cactus,* ABOVE, *may come from the Papago Indian word "opun" for this food plant shown growing with a common phacelia* (Phacelia sp.) *in Anza-Borrego State Park.*

off-road vehicles, open-pit leach mining, energy development, agriculture, and suburban sprawl have left indelible marks on California's deserts. Listing of the desert tortoise as an endangered species by the federal government has stimulated efforts in the Mojave Desert to develop large-scale regional plans to conserve biodiversity of both plants and animals.

In 1994, President Clinton signed into law the California Desert Protection Act, the largest wilderness protection bill ever for the lower forty-eight states. Under this law, 3.6 million acres of California desert are designated wilderness, administered primarily by the Bureau of Land Management, and three million acres are added to the National Park Service system. The long-term management changes that may ultimately arise from this legislation may be our best hope for the conservation of species and ecosystems of the California deserts.

Robert F. Thorne

WHITE MOUNTAINS

With the floor of Owens Valley more than 10,000 feet below and only a few miles to the west, the White Mountains exhibit some of the greatest extremes in relief in North America. They also contain the highest point in Nevada, 13,140-foot Boundary Peak, the third highest peak in California, 14,246-foot White Mountain Peak, and the oldest known living trees in the world, the bristlecone pine (*Pinus longaeva*).

Separated from the Inyo Mountains to the south by Deep Springs Valley and the 7,200-foot Westgard Pass, the White Mountains form the extreme southwestern corner of the Great Basin, the sagebrush country that extends across Nevada into Utah and to which most White Mountain plants are related.

The presence and evolution of many of these rare species result from past climate changes and the isolation of the White Mountains from nearby ranges, as well as from the more recent adaptation of some genera of plants to specialized habitats such as wetlands and chemically harsh soils.

The rarity of some populations of White Mountains plants makes them vulnerable to extinction by natural events such as years of climatic extremes, geologic catastrophes, and random die-offs.

Livestock grazing is a common activity in the White Mountains and poses a threat to rare plants and vegetation if not carefully managed. Water diversions, mining, off-road vehicles, loss of pollinating insects, and introduction of invasive exotic plants all threaten native plants in some areas of the White Mountains. The majority of land in the White Mountains is owned by the U.S. Forest

ANCIENT BRISTLECONE PINE

This picturesque, twisted pine occurs in small, isolated populations scattered across the higher mountain ranges of eastern California, Nevada, and western Utah. Often growing with the more common limber pine (*Pinus flexilis*), the five-needled bristlecone pine is actually more closely related to foxtail pine (*P. balfouriana*), also found in California, and is distinguishable by its shorter needles and reddish brown cones.

Bristlecone pines seem to live longest in severe habitats. Perhaps because they do not tolerate competition from other plants, trees are found most often on limestone or dolomite soils and rock outcrops. They survive mainly in the subalpine elevation zone between 9,500 and 11,500 feet where they experience extremes of heat and cold and

lack of moisture. In this harsh landscape, ancient bristlecone pines are very slow growing. The thin, poor soil, short growing season, and windy, dry climate result in a dense, resinous wood resistant to decay, disease, and breakage.

Because of their longevity and slow rate of decay, ancient bristlecone pines have been useful to scientists in reconstructing past climatic conditions. The width and density of each annual growth ring are determined by average temperatures, moisture, and other conditions during the year. By matching and overlapping the ring patterns of live trees with those found in dead wood, tree ring scientists have reconstructed an unbroken climatic record extending back some 8,700 years.

James D. Morefield

Service, which has set aside a Bristlecone Pine Preserve. There are smaller holdings owned by the Bureau of Land Management at lower elevations, and some private ownership.

James D. Morefield

The White Mountains, ABOVE, with their venerable bristlecone pines (Pinus longaeva), face the high Sierra across the Owens Valley, 10,000 feet below. • The diverse plant life of the White Mountains is a reflection of the diversity of rock and soil types and climatic zones. More than 930 native plants, or about one-sixth of California's flora, have been found here. Claret cup (Echinocereus triglochidiatus), FAR LEFT, is but one of two members of its group in California, and is perhaps our only truly red-flowered cactus. • Elk thistle (Cirsium scariosum), LEFT, is broadly distributed in the mountains of California and the West.

The ancient bristlecone pine (Pinus longaeva), ABOVE RIGHT, is the earth's oldest known living tree, with one specimen in the White Mountains dated at more than 4,600 years old. The oldest trees are characterized by broad bases, short stature, and large golden masses of gnarled and twisted dead wood with a few thin strips of bark leading to live needles. They tend to occur on the most exposed sites where growth is slowest. • As are the trunks, exposed roots are mostly dead wood, ABOVE.

OWENS VALLEY

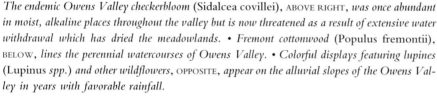

Owens Valley is a narrow trough between two of the highest mountain ranges in the continental United States, both reaching above 14,000 feet. Its west wall is the spectacular snowy Sierra Nevada, and its east wall the desertic White-Inyo Range.

Owens Valley, thousands of feet deep in alluvium, is about eighty-five miles long and as little as five miles wide. The Owens River originates in Mono County and, before its diversion to Los Angeles in 1913, it meandered through the valley to Owens Lake. The first European explorers to arrive in Owens Valley found a large, bluish body of water about seventeen miles long. Owens Lake began shrinking as early settlers diverted the streams that fed it. With continuous diversion, the lake basin is now a barren white expanse of alkaline deposits.

While the Sierra Nevada receives substantial precipitation, Owens Valley and the mountains eastward are in its rainshadow. The annual average precipitation for the southern part of the valley is five inches; the northern portion receives slightly more.

The City of Los Angeles owns most of the valley floor, much of which is grazed by livestock. The Bureau of Land Management and the Department of Fish and Game also own land on which they restrict grazing in the most sensitive areas.

This has been a difficult century for the water-dependent plants of Owens Valley because of the politics of water diversion. But there is hope. The Los Angeles Department of Water and Power and Inyo County are striving for a long-term water agreement that would, among other things, prevent excessive groundwater pumping and protect sensitive plant species. The U.S. Fish and Wildlife Service is leading an interagency planning effort with the California Department of Fish and Game, Bureau of Land Management, and the U.S. Forest Service to protect and recover rare and sensitive plants and animals in aquatic, riparian, and alkali marsh habitats of Owens Valley.

Mary DeDecker

The endemic Owens Valley checkerbloom (Sidalcea covillei), ABOVE RIGHT, *was once abundant in moist, alkaline places throughout the valley but is now threatened as a result of extensive water withdrawal which has dried the meadowlands.* • *Fremont cottonwood* (Populus fremontii), BELOW, *lines the perennial watercourses of Owens Valley.* • *Colorful displays featuring lupines* (Lupinus spp.) *and other wildflowers*, OPPOSITE, *appear on the alluvial slopes of the Owens Valley in years with favorable rainfall.*

FISH SLOUGH

Fish Slough, a unique and isolated desert wetland, is located in Owens Valley in a transitional area between the Great Basin and the Mojave Desert. Water flowing from some of the last natural springs in the Owens Valley creates extensive year-round and seasonal wetlands at Fish Slough before draining southward into the Owens River near Bishop.

Moisture and soil salinity levels vary throughout the Fish Slough area, creating a variety of habitats. There are ponds, springs, seasonally flooded alkali meadows, salt-encrusted alkali flats, and mosaics of alkali and alluvial scrub at slightly higher elevations. Partly because of low soil salinities, Fish Slough's rich flora includes 126 species of wetland plants. Today Fish Slough also serves as one of the only refuges for the endangered Owens Valley pupfish (*Cyprinodon radiosus*), a tiny fish that once ranged through the Owens Valley river system from Fish Slough south to Lone Pine in Inyo County. By the 1930s, agricultural interests and the City of Los Angeles had diverted much of the valley water flowing into the Owens River, and most of the pupfish's habitat was lost.

The Bureau of Land Management established Fish Slough as an Area of Critical Environmental Concern in 1985. It manages the area jointly with other private, state, and federal agencies and the University of California Natural Reserve System. Portions of Fish Slough are also protected as a state Ecological Reserve, a native fish sanctuary, and by designation as a National Natural Landmark. Management of this unique area is complicated because of conflict between grazing and recreational interests and policies established to protect endangered species and wetlands.

Wayne R. Ferren, Jr.

Desert wetlands such as those at Fish Slough, LEFT, *support species not found in more arid desert areas. Wetlands are vulnerable to water diversion, grazing, and thirsty non-native plants; consequently, most California desert wetlands are badly degraded.* • *Among the rare plants at Fish Slough are Fish Slough milk-vetch* (Astragalus lentiginosus *var.* piscinensis), TOP RIGHT, *known only from the alkali wetlands at Fish Slough, and the beautiful Inyo County star-tulip* (Calochortus excavatus), ABOVE RIGHT. • *Looking across Fish Slough, early morning light on the White Mountains in Owens Valley,* BELOW, *creates a world of changing light and shadow.*

Unlike many dune systems, the Eureka Dunes, ABOVE, are not in a rain shadow. Situated immediately west of the precipitous Last Chance Range, the great mass of sand acts as an above-ground aquifer. As rainwater slowly percolates through the sand mountain, it sustains a rich flora around its perimeter, even in the driest times of year. • Eureka Valley dune grass (Swallenia alexandrae), LEFT, thrives on pure sand slopes and reproduces from seeds or underground stems in deep sands. • This fascinating plant of ancient lineage, RIGHT, is known as July gold (Dedeck-era eurekensis). It bears the name of its discoverer, Mary DeDecker, who first found it in can-yons near the Eureka Dunes in the 1970s. It flowers in mid-summer, when daytime tempera-tures often reach 115 degrees F. • The large white flowers of the Eureka Dunes evening primrose (Oenothera californica ssp. eurekensis), ABOVE RIGHT, open in the evening and wilt when touched by the intense morning sun. This evening primrose has adapted to dune life through its ability to root at the nodes when the stems become covered with sand.

EUREKA VALLEY

Eureka Valley, an arc-shaped basin just west of the northern Death Valley drainage, is best known for its striking sand dunes. The Eureka Dunes, at the valley's southern end, are dominated by an impressive sand mountain nearly 700 feet high. The pale forms of these dunes are a dramatic contrast to the richly banded limestone and dolomite Last Chance Mountains in the east. This dry alluvial valley is bounded on the south by the Saline Range, and to the west and north by the Inyo Mountains. The dry creosote bush-covered valley floor drains from 3,500 feet at the north end to a sink at its southeast corner where the elevation is 2,920 feet.

Three endemic plants grow only in the deep sands of the Eureka Dunes. A foot-tall Eureka Valley dune grass (*Swallenia alexandrae*) forms large clumps and grows nearly to the top of the highest dune. Eureka Dunes evening primrose (*Oenothera californica* ssp. *eurekensis*) has grayish leaves and large, four-petaled, white flowers. It is found along the base of the dunes. Shining milk-vetch (*Astragalus lentiginosus* var. *micans*) is abundant along the sandy apron surrounding the dunes.

All forms of life on the dunes were seriously threatened for many years when they were a favorite site for off-road vehicle activity. The Bureau of Land Management eventually made the Eureka Dunes an area of critical environmental concern, and closed the area to motorized vehicles. In 1983 these remarkable dunes were designated a national natural landmark. In 1994, following the signing of the Desert Protection Act, the Eureka Dunes were added to Death Valley National Park.

Today, the plants and animals are recovering, but an aggressive and prolific barbwire Russian thistle (*Salsola paulsenii*), which became established during the soil-disturbing period of off-road vehicle abuse, now threatens to crowd out native species, particularly the endangered evening primrose.

Mary DeDecker

205

Inyo Mountains

Travelers viewing the Inyo Mountains may consider them featureless and barren, but those who know them intimately recognize the wealth of plants growing in their steep and rugged canyons. Pre-Cambrian formations are exposed above 10,000 feet, while granite outcrops provide scenic contrast. The range's highest peak at 11,123 feet, Waucoba Mountain is mostly siltstone and lacks the interesting flora that occurs over much of the range on exposed dolomite. There are numerous seeps and springs, and several small streams on the east side of the Inyos above Saline Valley.

The Inyos form the southern half of the White-Inyo Mountain Range. Although not as high as the White Mountains, they, along with the Panamints, are otherwise the highest of the desert mountains. The dividing line between the White and the Inyo mountains extends approximately east of Big Pine through the scenic Westgard Pass.

Bristlecone pine (*Pinus longaeva*) forests occur along most of the crest of the Inyo Mountains. Limber pines (*Pinus flexilis*) are also found at the higher elevations. Vast pinyon-juniper woodlands clothe the slopes and flats below. An unusually rich flora is found within these forests and woodlands. Examples of the many plants endemic to the Inyos are bristlecone cryptantha (*Cryptantha roosiorum*), known from only two populations; Jaeger's caulostramina (*Caulostramina jaegeri*), known from only about five populations; and broad-shouldered milk-vetch (*Astragalus cimae* var. *sufflatus*), found only in rock crevices and on cliffs here.

Scarcity of water and lack of access roads have limited both development and large-scale mining in the Inyo Mountains, though a fortune in silver was extracted from the Cerro Gordo region. Nevertheless, mining, grazing, firewood cutting, and illegal access roads threaten sensitive habitats. Feral goats, moved from San Clemente Island because of their destructive impact on the vegetation there, are now breeding in the rugged canyons of the Cerro Gordo, where they cannot be controlled. In addition, an ongoing proliferation of jeep trails and off-road vehicle routes degrades areas of wilderness quality.

Despite these incursions, the Inyo Mountains remain relatively untarnished. They are primarily within the Inyo National Forest; most of the Inyos outside Forest Service jurisdiction are now managed as wilderness by the Bureau of Land Management. Wilderness designation will help ensure that the unusual habitat values of the Inyo Mountains are preserved for generations to come.

Mary DeDecker

Dramatic rock formations, OPPOSITE, *support the single-leaf pinyon* (Pinus monophylla) *forest so abundant in the Inyo range; curl-leaf mountain-mahogany* (Cercocarpus ledifolius *var.* intermontanus), *in the foreground, is used locally to cure smoked meats.* • *In early summer desert mariposa lily* (Calochortus kennedyi *var.* kennedyi), LEFT, *adds spots of bright color to the lower slopes of the mountains.* • *The enormous genus* Astragalus, *with about 2,000 species worldwide, boasts few representatives as strikingly colored as the scarlet milk-vetch* (Astragalus coccineus), ABOVE. *It is California's only red-flowered milk-vetch.* • *The lovely desert-aster* (Machaeranthera tortifolia), BELOW, *is a common dweller of rocky cliffs and desert flats.*

DEATH VALLEY

eath Valley National Park covers 3.4 million acres of the northern Mojave Desert's most extraordinary terrain. Not only are the celebrated dune systems, mountains, and valleys fantastic to behold, but the great topographic relief and soil diversity make Death Valley one of California's most important centers of plant endemism. As dry and inhospitable as Death Valley appears, more than 900 species of vascular plants, nineteen of which are endemic, are found only here.

Death Valley's severe gradations of elevation and salinity help create a complex physical and biological environment. The vertical rise of more than two miles from Badwater, at 282 feet below sea level, to the top of Telescope Peak, at 11,000 feet eleva-

tion, is one of the steepest gradients in the United States.

Plant associations reflecting eleva-

tional changes range from salt-tolerant assemblages dominated by iodine bush (*Allenrolfea occidentalis*) on the valley floor to subalpine woodland with limber pine (*Pinus flexilis*) and bristlecone pine (*P. longaeva*). In between, desert holly (*Atriplex hymenelytra*) and creosote bush (*Larrea tridentata*) appear quite evenly spaced on the lower fans as they compete for limited moisture. Intergrading with them are burro-weed (*Ambrosia dumosa*) and shadscale (*Atriplex confertifolia*) on the upper fans. Middle elevations are a transition between the Mojave and Great Basin desert vegetation types and are occupied by blackbush (*Coleogyne ramosissima*) and a series of mixed scrub assemblages—spiny hopsage (*Grayia spinosa*), Nevada ephedra (*Ephedra nevadensis*), and spiny menodora (*Menodora spinescens*). These merge into sagebrush scrub and pinyon-juniper woodland at upper elevations below the subalpine woodlands.

The greatest threats to the survival of some of Death Valley's rarest plants have been from mining, feral burros, and proposed groundwater diversions. Following the Desert Protection Act of 1994, Death Valley National Monument was designated a national park, and about ninety percent of the park is designated wilderness. It is expected that national park status will afford greater protections to the extraordinary botanical riches of Death Valley.

Peter G. Rowlands

The Panamint daisy (Enceliopsis covillei), ABOVE LEFT, *a well known and showy flower of the mountains bordering Death Valley, has beeen adopted by the California Native Plant Society for its logo.* • *Nearly forty percent of Death Valley's flora are annual herbs such as gravel-ghost* (Atrichoseris platyphylla), LEFT, *the only member of its genus.* • *In a very dry year not one desert-sunflower* (Geraea canescens), TOP, *can be found on Death Valley's floor.* • *While spectacular wildflower displays blanket valley floors and slopes in years of good rainfall, many of Death Valley's rare plants are limited to specialized soils such as limestone and dolomite. Rock midget* (Mimulus rupicola), ABOVE, *clings to limestone crevices in two Death Valley desert canyons.* • *Golden carpet* (Gilmania luteola), ABOVE RIGHT, *can be found in only five areas on barren gypsum-rich soils in Death Valley.* • *Purple mat* (Nama demissum), RIGHT, *is one of the more common annuals in Death Valley.*

RED ROCK CANYON

Ten million years ago, Red Rock Canyon, home to one of California's rarest plants, echoed with the sounds of mastodons, saber-toothed tigers, and other mammals now extinct. Often receiving less than seven inches of annual rainfall today, this region once recorded twice as much rainfall and supported moister habitats such as woodland, savanna, and chaparral. The canyon now contains the most complete and varied fossil record in California of animal life from the early Pliocene epoch.

Botanists believe that the Red Rock tarplant (*Hemizonia arida*) is a remnant species from the now vanished Pliocene woodland that once covered this part of the Mojave Desert. The tarplant has apparently survived for thousands of years despite the extinction of its woodland plant and animal cohorts, by becoming confined to the courses of ephemeral streams and seeps, the wetter microhabitats in today's otherwise arid Mojave Desert environment.

Following the influx of European

settlers into California, a new set of threats was introduced to the habitat of this survivor. For decades, the narrow canyon was a thoroughfare for tens of thousands of sheep and horses, profoundly degrading the canyon's fragile ecosystem. In the 1960s, an even more destructive activity emerged: Red Rock Canyon became a favorite playground for off-road vehicle enthusiasts. When botanists visited the area in 1966, the seemingly adaptable tarplant appeared to be doomed.

Fortunately for the tarplant and other plants and animals, Red Rock Canyon was made a state park in 1970, and today vehicle activity is restricted to designated areas. However, outside the park, parts of the tarplant habitat that occur on federal lands are still used by off-road enthusiasts. Scars from past mining activities and invasion of an aggressive non-native tamarisk (*Tamarix ramosissima*) pose other management problems for the tarplant. The state park has developed an ongoing effort to remove the water-greedy tamarisk. In addition, the park has a policy of revegetating mined and denuded areas to a more natural condition. Given a vigilant and caring public, the park will have the support it needs to manage for the continued existence of habitat for tarplant, which has survived millions of years of climate change and more than 100 years of humans at work and play.

Roxanne L. Bittman

Cliffs striped with brilliant pinks, reds, and grays of an almost unearthly quality, OPPOSITE, *give this picturesque desert canyon, lying at the western end of the El Paso Mountains in Kern County, its name.* • *In spring, the open creosote bush scrub is brightened by an understory of flowering annual plants. The abundant Parry's linanthus (Linanthus parryi),* LEFT, ABOVE, *can be white or lavender.* • *Mariposa lilies, both common (desert mariposa,* Calochortus kennedyi, TOP) *and rare (striped mariposa lily,* C. striatus, LEFT, BELOW), *occur at Red Rock Canyon.* • *The Asteraceae, represented here by the desert-aster (Machaeranthera tortifolia),* LEFT, *is the largest plant family in California with over 900 different types.* • *Charlotte's phacelia (Phacelia nashiana),* ABOVE MIDDLE, *is a rare plant that grows in desert washes and on gravelly soils.* • *Red Rock tarplant (Hemizonia arida),* ABOVE, *a small, yellow-flowered annual in the sunflower family, and probably a remnant of a now vanished Pliocene woodland, grows only at Red Rock Canyon.*

EASTERN MOJAVE DESERT

An area of great scenic beauty and special biological interest occurs in the eastern portion of the Mojave Desert in the northeastern corner of San Bernardino County. Here are included in a number of mountain ranges including the Granite, Providence, New York, and Mesquite mountains, Mid Hills, and Ivanpah, Clark Mountain, and Kingston ranges, as well as the Kelso Dunes. The Kingston Range has the largest area, 158,885 acres, and the highest range, the Clark Mountain Range, reaches an elevation of over 7,900 feet. The area is subject to the usual arid desert climate with hot, dry summers and cool, dry winters, but because of their elevation, the ranges are moister and cooler than the surrounding desert floor. The eastern Mojave Desert receives some summer rainfall, which diminishes westward from the Colo-rado River. Areas with large marine deposits of limestone and dolomite, along with gypsum-rich and copper-rich soils, have a special interest for the botanist, for here are found rare and unusual plant species.

The great variety of habitats resulting from varied substrates, elevations, slopes, and exposures give rise to diverse plant communities. Among the more notable of these are the white fir-pinyon woodlands of north-facing higher slopes and narrow canyons in the highest ranges; pinyon-juniper-oak scrub found below the pinyon-juniper woodlands between the Mid Hills and New York Mountains; Joshua tree grassy woodlands on the well drained soils of mesas and slopes between 3,500 and 4,500 feet; creosote bush scrub of well drained bajadas, flats, and basins below 4,600 feet; desert scrub on the limestone and dolomitic substrates; scrub on gypsum-rich soils in the Clark Mountain Range; desert oasis woodland around permanent springs; and streamside marsh in some of the larger granitic canyons with shallow water flowing part of the year.

Most of the area was designated the Mojave National Preserve by the California Desert Protection Act, passed October 31, 1994. Prior to the Desert Act the preserve was managed by the Bureau of Land Management (BLM), but is now administered by the National Park Service. In addition to cre-ating the Mojave National Preserve, the Act also designated the majority of the higher ranges of the area as wilderness, some of which are managed by the National Park Service and some by BLM.

Robert F. Thorne

Rock-nettle (Eucnide urens), LEFT, ABOVE, *is not a true nettle, but it does have stinging hairs, and it grows in rocky soils.* • *Joshua tree* (Yucca brevifolia), LEFT, *has the shortest leaves and the greatest height of all the yuccas.* • *The handsome blazing star* (Mentzelia involucrata), TOP, *is closely related to rock-nettle, a similarity revealed by their flowers.* • *The eastern Mojave,* ABOVE, *is now protected through the 1994 Desert Protection Act.* • *The widespread and glorious desert mallow* (Sphaeralcea ambigua), RIGHT, *is covered with star-shaped hairs that can irritate the human eye.*

ANZA-BORREGO DESERT STATE PARK

Anza-Borrego Desert State Park is California's largest state park, preserving some of the best examples of western Colorado Desert habitats within its 600,000 acres. The park extends from the eastern escarpment of the Laguna Mountains in San Diego County to the Fish Creek Mountains of western Imperial County and from the Santa Rosa Mountains of southern Riverside County south almost to the U.S.-Mexican border. Elevations rise to 6,193 feet in the northwest corner of the park and fall to fifteen feet above sea level near the boundary of the Salton Basin.

The slopes and floor of the western Colorado Desert are dominated by shrubs such as creosote bush (*Larrea tridentata*), brittlebush (*Encelia farinosa*), burro weed (*Ambrosia dumosa*), ocotillo (*Fouquieria splendens*), and several cactus species. In years of abundant rainfall, spring displays of annual flowers provide spectacular color on the desert floor and slopes. More than two dozen groves of California fan palms (*Washingtonia filifera*) are scattered through-

Woolly Indian paintbrush (Castilleja foliolosa), FAR LEFT, TOP, is frequent on dry brushy slopes and blooms early in the season. • The vegetation of Anza Borrego, LEFT, includes many gorgeous desert shrubs, none more welcome than brittlebush (Encelia farinosa) and red-flowered chuparosa (Justicia californica) (shown here in the foreground) with arid desert mountains beyond. • Dune evening primrose (Oenothera deltoides ssp. deltoides) and desert sand-verbena (Abronia villosa var. villosa), ABOVE LEFT, can be glorious in years of early winter rains. • Along with those of the ocotillo (Fouquieria splendens), the red flowers of chuparosa (Justicia californica), ABOVE RIGHT, sustain breeding populations of Costa's hummingbirds, while the blue flowers of Phacelia distans attract bees. • The fragrant desert sand-verbena, BELOW LEFT, belongs to the four o'clock family and is common on open sandy deserts. • Wild canterbury bell (Phacelia campanularia), BELOW RIGHT, a striking desert annual, can produce dermatitis on some people.

out the park, adjacent to springs, seeps, and riparian pools. Pinyon-juniper woodlands occur at higher elevations.

Many rare plant species occur within the park such as sand-loving Gander's cryptantha (*Cryptantha ganderi*), an annual found in the sandy outwash of Coyote Canyon, and the delicately-flowered Arizona carlowrightia (*Carlowrightia arizonica*), found only in Hellhole and Borrego Palm canyons. Almost all California populations of the rare elephant tree (*Bursera microphylla*) are protected within the park.

Invasive non-native plants increasingly threaten native plants. Tamarisk (*Tamarix ramosissima*) can dry up entire desert streams; African fountain grass (*Pennisetum setaceum*) is spreading unchecked along rocky slopes and streams. Increases in residential and resort development and agriculture have lowered the water table on the valley floor around Borrego Springs. Botanists believe this may be responsible for a dramatic die-off of honey mesquite (*Prosopis glandulosa* var. *torreyana*) in the area.

James C. Dice

ALGODONES DUNES

The Algodones Dunes is the largest dune system in California, extending approximately forty-five miles along the eastern edge of the Imperial Valley in southeastern Imperial County. Sometimes referred to as the Imperial Sand Hills, this sea of sand extends for about five miles into Mexico to the mouth of the Colorado River. In some places along its length the sandfield may be over five miles wide, but the average width is closer to three miles. The highest sand peaks rise to more

The flowering heads of the rare sand food (Pholisma sonorae), ABOVE, a peculiar root parasite, resemble small button mushrooms growing on the surface of the dunes. Its common name refers to the plant's unseen underground portion, a fleshy stem that can extend several feet below the surface, which was once a prized food of local Native Americans. • Desert-marigolds (Baileya spp.), LEFT, form a striking pattern against the hot sand. • Pholisma arenarium, ABOVE RIGHT, a close relative of sand food found growing in Anza Borrego State Park is parasitic on shrubby members of the sunflower family. • With an average annual rainfall of less than two inches and summer temperatures that often rise above 110 degreees F., the Algodones Dunes, FAR RIGHT, are one of the hottest and driest areas in North America. Vegetation is adapted to arid, unstable, shifting sands and deep water percolation.

than 300 feet above the surrounding desert floor.

The Algodones Dunes represent a massive accumulation of wind-blown sand believed to have been derived from beach deposits of the freshwater Pleistocene Lake Cahuilla, whose fossil shoreline can still be seen at the base of the Santa Rosa Mountains near the Riverside-Imperial county line. The dune system consists of overlapping curved sand ridges that lie at right angles to the length of the dune field and alternate with sandy-floored depressions or hollows.

The Algodones Dunes are home to an interesting assemblage of rare and endangered plants. Four of these—the yellow Wiggin's croton (*Croton wigginsii*), bright yellow Algodones Dunes sunflower (*Helianthus niveus* ssp. *tephrodes*), Peirson's milk-vetch (*Astragalus magdalenae* var. *peirsonii*), and the strange plant called sand food (*Pholisma sonorae*)—are found mostly within the dunes.

The California portion of the Algodones Dunes is managed by the Bureau of Land Management. Following passage of the 1994 Desert Protection Act, the portion of the dunes north of Highway 78 and extending to Mammoth Wash was designated the North Algodones Dunes Wilderness. Nearly seventy percent of the dune system (more than 118,000 acres) remains open to off-road vehicles, which pose a major threat to the survival of both common and rare plant species there.

James C. Dice

THE DESERT FAN PALM

Desert fan palm (*Washingtonia filifera*), sometimes called California fan palm, with its rich green crown, makes a striking appearance at the approximately 150 desert oases where it occurs. The vast majority of palm oases are in the Colorado Desert, where vegetation is typically small and dormant for most of the year.

The isolation of palm oases and the attractiveness to desert adventurers has contributed to their demise. Several palm oases along the San Andreas fault in California's Coachella Valley have been obliterated by off-road vehicles, target practice, and camping within the oasis environment. Other palm oases are threatened by the pumping of underground water supplies for human use—a practice that has eliminated dozens of trees at Macomber Palms in Riverside County and threatens many more in the Oasis of Mara in San Bernardino County. Although desert fan palms can withstand ground fires, vandals sometimes torch the palms themselves, and some palm oases have suffered from torching or uncontrolled wildfires so many times that younger, less resistant trees are being eliminated. Unfortunately, frequent wildfires are degrading oases at Forty-Nine Palms in Joshua Tree National Monument and at Bear Creek Palms near the town of La Quinta. Education is needed so that the public will value and protect remaining fan palm oases.

James W. Cornett

Desert fan palm (Washingtonia filifera), ABOVE, *sometimes called the California fan palm, is the only palm tree native to the western United States. It is also the largest plant occurring at most permanent seeps, springs, and streams where it occurs. Growing up to eighty feet tall and with a trunk nearly four feet across, the desert fan palm is the largest palm, by weight, native to North America. It is the most cold-tolerant palm species in the world and is the only member of the palm family whose leaves typically adhere to the entire length of the trunk throughout the tree's life. It is also the only plant species whose above-ground portion survives an oasis fire.*

TOTAL NUMBERS OF RARE AND ENDANGERED PLANTS BY COUNTY

County/ Island	CNPS Listed[1]	State/Federal Listed	County/ Island	CNPS Listed	State/Federal Listed
Alameda	68	9	San Benito	62	3
Alpine	20	0	San Bernardino	234	16
Amador	28	2	San Diego	211	28
Butte	80	4	San Francisco	32	8
Calaveras	26	1	San Joaquin	32	7
Colusa	55	5	San Luis Obispo	153	24
Contra Costa	67	9	San Mateo	59	12
Del Norte	117	3	Santa Barbara	91	14
El Dorado	46	6	Santa Clara	66	7
Fresno	118	14	Santa Cruz	59	11
Glenn	41	4	Shasta	99	3
Humboldt	111	6	Sierra	37	0
Imperial	53	3	Siskiyou	187	6
Inyo	159	10	Solano	47	5
Kern	122	10	Stanislaus	46	8
Kings	16	3	Sonoma	125	22
Lake	93	9	Sutter	6	1
Lassen	69	2	Tehama	87	6
Los Angeles	135	15	Trinity	118	2
Madera	47	7	Tulare	112	13
Marin	89	16	Tuolumne	69	4
Mariposa	58	7	Ventura	69	11
Mendocino	106	13	Yolo	29	3
Merced	46	7	Yuba	15	1
Modoc	66	1			
Mono	102	4	Anacapa Island	28	1
Monterey	139	18	San Clemente Isl.	56	7
Napa	89	8	San Miguel Isl.	24	2
Nevada	40	4	San Nicolas Isl.	24	3
Orange	69	9	Santa Barbara Isl.	17	3
Placer	37	2	Santa Catalina Isl.	51	3
Plumas	71	1	Santa Cruz Isl.	62	5
Riverside	159	14	Santa Rosa Isl.	54	1
Sacramento	21	5			

[1] Total CNPS Listed = 1A, 1B, 2, 3 and 4; the California Native Plant Society (CNPS) publishes their *Inventory of Rare and Endangered Vascular Plants of California* (Skinner and Pavlik, 1994) and defines their lists as follows:

    1A.   Presumed Extinct in California

    1B.   Rare or Endangered in California and Elsewhere

    2.   R/E in California, More Common Elsewhere

    3.   Need More Information

    4.   Plants of Limited Distribution

GLOSSARY

Alkali sink: Moist, undrained habitat where water remains standing much of the year

Alluvium: Clay, silt, sand, or similar detrital material deposited by running water

Anthropogenic rarity: A plant that has become rare due to human-produced environmental change

Arroyo: A watercourse in an arid area

Badlands: Region marked by little vegetation and intricate erosional sculpting

Basalt: A dark gray to black, dense to fine-grained igneous rock

Berm: Narrow shelf or ledge around the bottom of a slope

Botanical Area: Official designation of the U.S. Forest Service

Calcareous soil: chalky, rich in calcium

Calcium carbonate: Calcite, one of the commonest minerals, the principal constituent of limestone

Chaparral: Drought-tolerant scrub vegetation

Claypan: Compacted layer of clay, often causing seasonal water ponding

Colluvium: Rock fragments and soil accumulated at the foot of a slope

Dolomite: A limestone or marble rich in magnesium carbonate

Edaphic: Pertaining to, or influenced by, soil conditions

Endemic: A species occurring in specific area with restricted distribution

Endemism: Restricted to a specific locality, often to a certain soil type for plants

Fault-block mountain: A mountain formed by the combined processes of uplifting, faulting, and tilting

Fen: Open stretch of wet marshy land with base-rich water and build up of basic peat

Franciscan Formation: Assemblage of mostly sedimentary rocks formed in the Mesozoic era and comprising much of the Coast Ranges and rocks around San Francisco Bay

Gabbro: Rock with high iron and magnesium content found on ocean/continental margins, often underlying reddish, clay-like soil

Gypsum: Hydrated calcium sulfate, a common mineral

Igneous rock: Formed by cooling of molten volcanic magma on or near the Earth's surface

Mediterranean climate: characterized by wet winters and hot, dry summers

Mesic: requiring a moderate amount of moisture

Metamorphic rock: Formed by heat and pressure at depth changing the structure of preexisting rock

Metasedimentary: A sedimentary rock showing evidence of metamorphic change at a depth in the Earth's crust

Miocene: Geologic era between 25,000,000 and 5,000,000 years ago characterized by presence of grazing mammals

Paleoendemic: A narrowly distributed remnant of a past floral association

Plant community: A sub-category of vegetation type named for the dominant plant species (eg. ponderosa pine forest)

Pleistocene: Geologic era between 1,000,000 and 10,000 years ago

Pumice: A porous or spongy form of volcanic glass

Radiate: Growth around a center, usually symmetrical

Relict: A persistent remnant of an otherwise extinct life form

Rhyolite: An acidic granite made from volcanic lava

Riprap: A retaining wall of loose stones, usually found at the edge of a body of water, man-made

Sag pond: Body of water formed in a depression along a faultline

Sedimentary rock: Formed by the deposition of sediment

Senescent: Plant growth phase between maturity and death

Serpentine: California's state rock, a metamorphic magnesium silicate that weathers into toxic soils (high in magnesium and low in calcium) supporting rare and endemic plants.

Sink: A slight, low-lying desert depression containing a central playa or saline lake with no outlet, as where a desert stream comes to an end or disappears by evaporation

Substrate: Base on which an organism lives, usually the soil for a plant

Swale: Low-lying, often moist ground

Talus: Slope formed by accumulation of rock debris

Temperate: Where coldest winter is mild—above 32°F and below 64°F

Tule fog: Cold, dense fog trapped by temperature inversion. Occurs in Central Valley in winter, and named after the tule reed found alongside streams

Ultramafic: Substrate very low in silica and rich in magnesium and iron

Vascular: The system of channels for moving and storing moisture and minerals within a plant

Vegetation type: A category of plant cover named for the geographic location and dominant plant form (eg. Central Valley annual grassland)

Vernal pool: Seasonal pool, usually filled with water in winter and early spring, and dry the rest of the year

ORGANIZATIONS YOU CAN WORK WITH AND JOIN

California Native Plant Society

1722 J Street, Suite 17
Sacramento, CA 95814
916/447-2677

California Academy of Sciences

Botany Department
Golden Gate Park
San Francisco, CA 94118
415/750-7187

California Exotic Plant Pest Council (CalEPPC)

P.O. Box 15575
Sacramento, CA 95852
916/921-5911

National Audubon Society

Western Regional Office
555 Audubon Place
Sacramento, CA 95814
916/481-5332

The Nature Conservancy

201 Mission Street, 4th Floor
San Francisco, CA 94105
415/777-0487

Planning and Conservation League

926 J Street, Suite 621
Sacramento, CA 95814
916/444-8726

The Sierra Club

State Legislative Office
923 12th Street, Suite 200
Sacramento, CA 95814
916/557-1100

PUBLIC AGENCIES YOU CAN WORK WITH

California Department of Fish and Game (CDFG)

1416 Ninth Street
Sacramento, CA 95814
916/653-7664

Bureau of Land Management (BLM)

California State Office
2135 Butano Drive
Sacramento, CA 95825
916/979-2800

U.S. Forest Service (USFS)

Pacific Southwest Region
630 Sansome Street
San Francisco, CA 94111
415/705-2874

U.S. Fish and Wildlife Service (USFWS):

Carlsbad Field Office

2730 Loker Avenue West
Carlsbad, CA 92008
619/431-9624

Sacramento Field Office

3310 El Camino, Room 130
Sacramento, CA 95821
916/979-2710

Ventura Field Office

2493 Portola Road, Suite B
Ventura, CA 93003
805/644-1766

ADDITIONAL READINGS

Barbour, M., B. Pavlik, F. Drysdale, and S. Lindstrom. 1993. *California's Changing Landscapes: Diversity and Conservation of California Vegetation*. California Native Plant Society. Sacramento, CA. 244 pp.

Donley, M.W., S. Allan, P. Caro, C.P. Patton. 1979. *Atlas of California*. Academic Book Center, Portland, OR. 191 pp.

Elias, T.S. 1987. *Conservation and Management of Rare and Endangered Plants*. Proceedings of a Conference of the California Native Plant Society. Sacramento, CA. 630 pp.

Falk, D.A. and K.E. Holsinger. 1991. *Genetics and Conservation of Rare Plants*. Oxford University Press. New York, NY. 304 pp.

Falk, D.A., C.I. Millar, and M. Olwell. 1996. *Restoring Diversity: Strategies for Reintroduction of Endangered Plants*. Island Press, Washington, D.C. 505 pp.

Fiedler, P. and S. Jain (eds.) 1992. *Conservation Biology: The Theory and Practice of Nature Conservation, Preservation, and Management*. Chapman and Hall, N.Y., N.Y.

Given, D.R. 1994. *Principles and Practice of Plant Conservation*. Timber Press, Portland, OR. 292 pp.

Gleason, H.A., A. Cronquist. 1964. *The Natural Geography of Plants*. Columbia University Press. N.Y., N.Y. 420 pp.

Gunn, A.S. 1980. "Why Should We Care About Rare Species?" *Environmental Ethics* 2(1):17-37.

Hickman, J.C., ed. 1993. *The Jepson Manual: Higher Plants of California*. University of California Press. Berkeley, CA. 1400 pp.

Hill, R.B. 1986. *California Mountain Ranges*. No. 1 in the California Geographic Series. Falcon Press. Helena, MT. 119 pp.

Jensen, D.B., M. Torn and J. Harte. 1990. *In Our Own Hands: A Strategy for Conserving Biological Diversity in California*. California Policy Seminar, University of California, Berkeley. 184 pp.

Jones & Stokes Associates. 1987. *Sliding Toward Extinction: The State of California's Natural Heritage, 1987*. The California Nature Conservancy. San Francisco, CA. 105 pp.

Koopowitz, H. and H. Kaye. 1983. *Plant Extinction: A Global Crisis*. Stone Wall Press. Washington, D.C. 239 pp.

Kruckeberg, A.R. 1984. *California Serpentines: Flora, Vegetation, Geology, Soils, and Management Problems*. University of California Press. Berkeley, CA. 180pp.

Kruckeberg, A.R. and D. Rabinowitz. 1986. "Biological Aspects of Endemism in Higher Plants." *Annual Review of Ecology and Systematics* 16:447-479.

McMahan, L. 1980. "Legal Protection for Rare Plants." *American University Law Review* 29(3):515-569.

McMahan, L. 1987. "Rare Plant Conservation by State Governments." In, Elias, T. S. 1987. *Conservation and Management of Rare and Endangered Plants*. Proceedings of a Conference of the California Native Plant Society. California Native Plant Society. Sacramento. 23-31 pp.

Munz, P.A., and D.D. Keck. 1959. *A California Flora*. University of California Press. Berkeley, CA. 1681 pp. plus supplement.

Niehaus, T., and C.L. Ripper. 1976. *A Field Guide to Pacific States Wildflowers*. Houghton Mifflin Company. Boston. 432 pp.

Pavlik, B.M., P.C. Muick, S. Johnson, and M. Popper. 1991. *Oaks of California*. Cachuma Press. Los Olivos, CA. 184 pp.

Prance, G.T. and T.S. Elias (eds). 1977. *Extinction Is Forever: Threatened and Endangered Species of Plants in the Americas and their Significance in Ecosystems Today and in the Future*. New York Botanical Garden. Bronx. 437 pp.

Raven, P.H., and D.I. Axelrod. 1978. *Origin and Relationships of the California Flora*. University of California Publications in Botany 72:1-134.

Reveal, J.L. 1981. "The Concepts of Rarity and Population Threats in Plant Communities." In. L.E. Morse and M.S. Henifin (editors). *Rare Plant Conservation: Geographical Data Organization*. New York Botanical Garden. Bronx. 41-47 pp.

Shoenherr, A.A. ed. "Endangered Plant Communities of Southern California." *Southern California Botanists Special Publication #3*. 1989:1-112.

Skinner, M.W. and B.M. Pavlik. 1994. *California Native Plant Society's Inventory of Rare and Endangered Vascular Plants of California*. California Native Plant Society, Sacramento, CA.

Soule, M.E. (ed.). 1986. *Conservation Biology: The Science of Scarcity and Diversity*. Sinauer Associates. Sunderland, MA. 584 pp.

Stebbins, G.L. 1978a. "Why Are There So Many Rare Plants in California? I. Environmental Factors." *Fremontia* 5(4):6-10.

Stebbins, G.L. 1978b. "Why Are There So Many Rare Plants in California? II. Youth and Age of Species." *Fremontia* 6(1):17-20.

Stebbins, G.L. 1980. "Rarity of Plant Species: A Synthetic Viewpoint." *Rhodora* 82:77-86.

Stebbins, G.L. 1986. "Rare Plants in California's National Forests: Their Scientific Value and Conservation." *Fremontia* 13(4):9-12.

Stebbins, G.L. and J. Major. 1965. "Endemism and Speciation in the California Flora." *Ecological Monographs* 35:1-35.

Synge, H. (ed.). 1981. *The Biological Aspects of Rare Plant Conservation*. John Wiley & Sons. New York, NY. 558 pp.

INDEX

Note: Page numbers in italic type indicate references to photographs. If a photograph appears on a different page than its caption, the page number given in the index is that of the photograph, not the caption.

AUTHOR ADDRESSES

Walt Anderson
Biology Department
Prescott College
4201 N. Covina Circle
Prescott Valley, AZ 86314

Dr. Michael Baad
Department of Biology
California State University
Sacramento, CA 95819-2694

Dr. Bruce C. Baldwin
Jepson Herbarium
University of California
Berkeley, CA 94720

Dr. Michael Barbour
Department of Environmen-
tal Horticulture
University of California
Davis, CA 95616

Dr. Ellen T. Bauder
Biology Department
San Diego State University
San Diego, CA 92182

R. Mitchel Beauchamp
Pacific Southwest Ecological
Services
P. O. Box 985
National City, CA 92050

Ken Berg
USFWS, Carlsbad Field
Office
2730 Loker Avenue West
Carlsbad, CA 92008

Roxanne L. Bittman
CDFG—Wildlife Data
Analysis
1416 Ninth Street
Sacramento, CA 95814

Dr. Mary Bowerman
Jepson Herbarium
University of California
Berkeley, CA 94720-2465

Dr. Peter Bowler
Museum of Systematic
Biology
University of California
Irvine, CA 92717-2525

Dr. Steven Boyd
Rancho Santa Ana Botanic
Garden
1500 North College Avenue
Claremont, CA 91711-3101

Angelika Brinkmann-Busi
2141 W. 35th Street
San Pedro, CA 90732

John W. Brown
USDA Systematics
Entomology Laboratory
c/o Smithsonian Institution
National Museum of Natural
History
Washington, D.C. 20560

Julie Carville
14582 Alderwood Way
Nevada City, CA 95959

Dr. David Chipping
1530 Bayview Heights Drive
Los Osos, CA 93402

Ronilee Clark
DPR—Southern Service
Center
8885 Rio San Diego Drive,
Suite 270
San Diego, CA 92108

Susan Cochrane
CDFG—Conservation
Education
1416 Ninth Street
Sacramento, CA 95814

James Cornett
Palm Springs Desert Museum
101 Museum Drive
Palm Springs, CA 92263

Susan D'Alcamo
202 Ridge Road
Sonora, CA 95370

Karen Danielsen
Great Lakes Indian Fish and
Wildlife Commission
P.O. Box 9
Odanah, WI 54861

Mary DeDecker
(Deceased)

James C. Dice
CDFG—Region 5
P.O. Box 2537
Borrego Springs, CA 92004

Dr. Stephen Edwards
Regional Parks Botanic
Garden
Tilden Regional Park
Berkeley, CA 94708

Anni Eicher
Box 527
Arcata, CA 95521

Dr. Barbara Ertter
UC and Jepson Herbaria
University of California
Berkeley, CA 94720-2465

Wayne R. Ferren, Jr.
67 Main Street
Vicentown, NJ 08088

Dr. Peggy Fiedler
BBL Sciences/Ecosystem
Science and Restoration
Services
2033 North Main Street,
#340
Walnut Creek, CA 94596

Ida Geary
51 Park Avenue
Mill Valley, CA 94941

Dr. Thomas Griggs
The Nature Conservancy
1658 Inghram Road
Corning, CA 96021

Betty Lovell Guggolz
CNPS Milo Baker Chapter
1123 Palomino Road
Cloverdale, CA 95425

Jennie Haas
USFS—Stanislaus National
Forest
24525 Highway 120
Groveland, CA 95321

Dr. J. Robert Haller
Santa Barbara Botanic
Garden
1212 Mission Canyon Road
Santa Barbara, CA 93106

Nancy and William Harnach
P.O. Box 28
Sattley, CA 96124

Deborah Hillyard
P.O. Box 1388
Morro Bay, CA 93445

Dr. Robert Holland
3371 Ayers Holmes Road
Auburn, CA 95603

Lisa Hoover
USFS—Six Rivers National
Forest
1330 Bayshore Way
Eureka, CA 95501

Ann Howald
210 Chestnut Avenue
Sonoma, CA 95476

Diane Ikeda
CDFG—Habitat
Conservation Planning
1416 Ninth Street
Sacramento, CA 95814

David Imper
CNPS—North Coast
Chapter
4612 Lentell Road
Euereka, CA 95501

James D. Jokerst
(deceased)

Dr. Steve Junak
Santa Barbara Botanic
Garden
1212 Mission Canyon Road
Santa Barbara, CA 93105

Dr. Glenn Keator
1455 Catherine Dr.
Berkeley, CA 94702

Dr. Todd Keeler-Wolf
CDFG—Natural Heritage
Division
1416 Ninth Street
Sacramento, CA 95814

Dr. David Keil
Department of Biological
Sciences
California Polytechnic State
University
San Luis Obispo, CA 93407

Tim Krantz
P.O. Box 33
Angelus Oaks, CA 92305

Dr. Arthur Kruckeberg
Department of Botany
University of Washington
Seattle, WA 98195

Dr. Earl Lathrop
Department of Biology
LaSierra University
4700 Pierce Street
Riverside, CA 92515-8247

Dr. Richard Lis
CDFG—Region 1
601 Locust Street
Redding, CA 96001

Mary Ann Matthews
P.O. Box 381
Carmel Valley, CA 93924-
0381

Stephen McCabe
205 Morningside Drive
Ben Lomond, CA 95005

Dr. Niall F. McCarten
Jones & Stokes Associates
2600 V Street, Suite 100
Sacramento, CA 96818-1914

Dr. Elizabeth McClintock
(Deceased)

Dr. Malcolm McLeod
Department of Biological
Sciences
California Polytechnic State
University
San Luis Obispo, CA 93407

Joseph Medeiros
Biology Department
Sierra College
5000 Rocklin Road
Rocklin, CA 95677-3397

James Morefield
Nevada Natural Heritage
Program
1550 E. College Parkway,
Suite 145
Carson City, NV 89710

Sandra Morey
CDFG—Habitat
Conservation Planning
1416 Ninth Street
Sacramento, CA 95814

Dr. Pamela Muick
2660 Gulf Drive
Fairfield, CA 94533

Dr. David Keil
Department of Biological
Sciences
San Jose State University
San Jose, CA 95125

Dr. Rodney Myatt
Department of Biological
Sciences
San Jose State University
San Jose, CA 95125

Nicole Nedeff
P.O. Box 1525
Carmel, CA 93924

Julie Kierstead Nelson
USFS—Shasta-Trinity
National Forest
2400 Washington Avenue
Redding, CA 96001

Virginia Norris
(deceased)

Dr. John O'Leary
Department of Geography
San Diego State University
San Diego, CA 92182-0381

Thomas Oberbauer
San Diego County Planning
Department
5201 Ruffin Road
San Diego, CA 92123

Dr. Robert Ornduff
(Deceased)

Dr. Thomas V. Parker
Department of Biological
Sciences
San Francisco State
University
San Francisco, CA 94132

Dr. Bruce Pavlik
Department of Biology
Mills College
Oakland, CA 94613

Dr. Ralph Philbrick
29 San Marcos Trout Club
Santa Barbara, CA 93105

Andrea Pickart
Lanphere Dunes Preserve
6800 Lanphere Road
Arcata, CA 95521

Barry Prigge
Mildred Mathias Botanical
Garden
405 Hilgard Avenue
Los Angeles, CA 90024-1606

Craig Reiser
(Deceased)

Fred Roberts, Jr.
USFWS—Carlsbad Field
Office
2730 Loker Avenue West
Carlsbad, CA 92008

Robert Rogers
USFS—Sequoia National
Forest
900 West Grand Avenue
Porterville, CA 93527

Dr. Edward Ross
Entomology Department
California Academy of
  Sciences
Golden Gate Park
San Francisco, CA 94118

Dr. Peter G. Rowlands
Northern Arizona University
Box 5614
Flagstaff, AZ 86011-5614

Cynthia Roye
DPR Resource Management
  Division
P.O. Box 942896
Sacramento, CA 94296-0001

Connie Rutherford
USFWS—Ventura Field
  Office
2493 Portola Road, Suite B
Ventura, CA 93003

Dr. John Sawyer
Department of Biological
  Sciences
Humboldt State University
Arcata, CA 95521

Dr. Robert Schlising
Department of Biological
  Sciences
California State University
Chico, CA 95929-0515

Gary Schoolcraft
BLM—Susanville District
2950 Riverside Drive
Susanville, CA 96130

James Shevock
Californian Cooperative
  Ecosystem Studies Unit
  (CA-CESU)
Department of Environmen-
  tal Science, Policy, and
  Management
University of California
337 Mulford Hall
Berkeley, CA 94720-3114

Teresa Sholars
Biology Department
College of the Redwoods
1211 Del Mar Drive
Ft. Bragg, CA 95437

David Showers
CDFG—Environmental
  Services Division
1416 Ninth Street
Sacramento, CA 95814

Mary Ann Showers
Division of Mines and
  Geology
801 K Street, MS 12-32
Sacramento, CA 95814-3531

Jacob Sigg
338 Ortega Street
San Francisco, CA 94122

Dr. Mark Skinner
USDA—National Plant Data
  Center
Box 74490
Baton Rouge, LA 70874

Susan Sommers
CNPS—Santa Clara Valley
  Chapter
887 Roble Avenue #3
Menlo Park, CA 94025

Constance Spenger
1318 East Glenwood Avenue
Fullerton, CA 92631

Dr. G. Ledyard Stebbins
(Deceased)

Dr. John Stebbins
Department of Biology
California State University
Fresno, CA 93740

Dr. Dean Taylor
UC and Jepson Herbaria
University of California
Berkeley, CA 94720-2465

Tim Thomas
8448 SVL Box
Victorville, CA 92392

Dr. Robert F. Thorne
Rancho Santa Ana Botanic
  Garden
1500 North College Avenue
Claremont, CA 91711

Dr. Robbin Thorp
Department of Entomology
University of California
Davis, CA 95616

William Tippets
CDFG—Environmental
  Services Division
4949 Viewridge Avenue
San Diego, CA 92123

Valerie Whitworth
Box 757
Winters, CA 95694

Peter Van Solen
Cal Flora Nursery
2990 Somers Street
Fulton, CA 95439

Roy van de Hoek
Catalina Interpretive Center
Los Angeles County Parks
  and Recreation Depart-
  ment
Avalon, CA 90704-2072

Dr. Delbert Weins
Department of Biology
University of Utah
201 Biology Building
Salt Lake City, UT 84112

Howard Wier
Dudek & Associates, Inc.
605 Third Street
Encinitas, CA 92024

Robert Wunner
951 18th Street
Arcata, CA 95521

PHOTOGRAPHER ADDRESSES

Dr. Frank Almeda
Botany Department
California Academy of
  Sciences
Golden Gate Park
San Francisco, CA 94118

Walt Anderson
Biology Department
Prescott College
4201 N. Covina Circle
Prescott Valley, AZ 86314

Wayne Armstrong
Palomar College—Life
  Sciences
1140 W. Mission Road
San Marcos, CA 92069

Marianne Austin-McDermon
P.O. Box 1244
Sonoma, CA 95476

Dr. Bruce C. Baldwin
Jepson Herbarium
University of California
Berkeley, CA 94720

Dr. Gregory Ballmer
Department of Entomology
University of California
Riverside, CA 92521-0314

Frank Balthis
Nature's Design
P.O. Box 255
Davenport, CA 95017

Dr. Michael Barbour
Department of Environmen-
  tal Horticulture
University of California
Davis, CA 95616

Ken Berg
USFWS, Carlsbad Field
  Office
2730 Loker Avenue West
Carlsbad, CA 92008

Roxanne L. Bittman
CDFG—Natural Heritage
  Division
1416 Ninth Street
Sacramento, CA 95814

Tupper Ansel Blake
Wildlife Photography
P.O. Box 152
Inverness, CA 94937

Jan Briggs
75 Drake Court
Walnut Creek, CA 94596

Angelika Brinkmann-Busi
CNPS—South Coast
  Chapter
2141 W. 35th Street
San Pedro, CA 90732

David Cavagnaro
1575 Manawa Road
Rt. 2
Decorah, IA 52101

Dan Cheatham
3462 Rossmoor Pkwy #1
Walnut Creek CA 94595

Dr. George Clark
(deceased)

Susan Cochrane
CDFG—Natural Heritage
  Division
1416 Ninth Street
Sacramento, CA 95814

Ed Cooper
3800 Grove Street
Sonoma, CA 95476

James Cornett
Palm Spring Desert Museum
101 Museum Drive
Palm Springs, CA 92263

Buff & Gerald Corsi
Focus on Nature, Inc.
5570 Inverness Avenue
Santa Rosa, CA 95404

Ken W. Davis
5637 Cypress Point Dr.
Citrus Heights, CA 95610

Tommy Dodson
Edge of Eden
4470 Sunset Blvd. #278
Hollywood, CA 90027

Don Eastman
Scenic & Nature Photogra-
  phy
17532 NW Shadyfir Loop
Beaverton, OR 97006

Ed Ely
2683 42nd Avenue
San Francisco, CA 94116

Bill Evarts Photography
1831 Lendee Dr.
Escondido, CA 92025

Phyllis Faber
212 Del Casa Drive
Mill Valley, CA 94941

Gary Fellers
NPS—Point Reyes National
  Seashore
Pt. Reyes, CA 94956-9799

William and Wilma Follette
1 Harrison Street
Sausalito, CA 94965

Ellen Frank
1350 Peach Street,
Martinez, CA 94553

John Game
1155 Spruce Street
Berkeley, CA 94707

Dede Gilman
(deceased)

Michael Graf
2823 Palm Court
Berkeley, CA 94705

Mary Ann Griggs
1658 Inghram Road
Corning, CA 96021

Robert Gustafson
Natural History Museum Life
  Sciences
900 Exposition Blvd.
Los Angeles, CA 90007

Jennie Haas
USFS—Stanislaus National
Forest
24525 Highway 120
Groveland, CA 95321

Dr. J. Robert Haller
Santa Barbara Botanic
Garden
1212 Mission Canyon Road
Santa Barbara, CA 93106

Jessie Harris
Flower & Nature Photography
4401 W. Street, N.W.
Washington, D.C. 20007-1133

Art Hayler
(Deceased)

Dr. Lawrence Heckard
(deceased)

Richard Herrman Photography
12545 Mustang Drive
Poway, CA 92064

Deborah Hillyard
P.O. Box 1388
Morro Bay, CA 93445

Saxon Holt
P.O. Box 1826
Novato, CA 94948

David Imper
CNPS—North Coast
Chapter
4612 Lentell Road
Eureka, CA 95501

Paul Johnson
P.O. Box 1555
Borrego Springs, CA 92004

Treve Johnson
509 Carmel Avenue
Albany, CA 94706-1412

James D. Jokerst
(deceased)

Steve Junak
Santa Barbara Botanic
Garden
1212 Mission Canyon Road
Santa Barbara, CA 93105

Dr. Todd Keeler-Wolf
CDFG—Natural Heritage
Division
1416 Ninth Street
Sacramento, CA 95814

Dr. David Keil
Department of Biological
Sciences
California Polytechnic State
University
San Luis Obispo, CA 93407

Tim Krantz
18957 Patton Drive
Castro Valley, CA 94546-3134

Craig Lippus
Nature Photography
5025 Northlawn Drive
San Jose, CA 95130

Lynn Lozier
The Nature Conservancy
201 Mission Street
San Francisco, CA 94105

Harold Malde
842 Grant Plance
Boulder, CA 80302

Stuart McKelvey
(Deceased)

Dr. Malcolm McLeod
Department of Biological
Sciences
California Polytechnic State
University
San Luis Obispo, CA 93407

Joseph Medeiros
Biology Department
Sierra College
5000 Rocklin Road
Rocklin, CA 95677-3397

Dr. Constance Millar
USFS
800 Buchanan Street
Albany, CA 94710

Steve Montgomery
6389 Camanito Del Pastel
San Diego, CA 92111

Nicole Nedeff
P.O. Box 1525
Carmel, CA 93924

Thomas Oberbauer
3437 Trumbull Street
San Diego, CA 92106

Bart O'Brien
Rancho Santa Ana Botanic
Garden
WHY NO ADDRESS
HERE???

Jo-Ann Ordano
P.O. Box 591291
San Francisco, CA 94159

Ron Parsons
2704 Hillside Drive
Burlingame, CA 94010

Dr. Bruce Pavlik
Department of Biology
Mills College
Oakland, CA 94613

B. "Moose" Peterson
Wildlife Research Photography
P.O. Box 30694
Santa Barbara, CA 93130

Andrea Pickart
Lanphere Dunes Preserve
6800 Lanphere Road
Arcata, CA 95521

Ron Pickup
P.O. Box 62
Soulsby, CA 95372

Roger Raiche
12440 Occidental Rd.
Sebastopol, CA 95472

Joan Rosen Photo
61 Lee Street
Mill Valley, CA 94941

Dr. Edward Ross
Entomology Department
California Academy of
Sciences
Golden Gate Park
San Francisco, CA 94118

Galen Rowell
(deceased)

David Sanger
920 Evelyn Avenue
Albany, CA 94706

Dr. Robert Schlising
Department of Biological
Sciences
California State University
Chico, CA 95924-0515

Dr. Larry Serpa
The Nature Conservancy
201 Mission Street
San Francisco, CA 94105

James Shevock
NPS—Pacific West Region
600 Harrison Street
San Francisco, CA 94107

Dr. Mark Skinner
USDA—National Plant Data
Center
Box 74490
Baton Rouge, LA 70874

Linda Smith
P.O. Box 22124
Carmel, CA 93922

Dr. James Smith, Jr.
Department of Biological
Sciences
Humboldt State University
Arcata, CA 95521

Jon Mark Stewart
8020 Dark Mesa, NW, Apt.
C18
Albuquerque, NM 87120

Kathleen Stockwell
PO Box 955
Trabuco Canyon, CA 92678

Dan Suzio
P.O. Box 5803
Berkeley, CA 94705

Dale Thomas
Rt. 2, Box 321
Mason, WI 54856

Tim Thomas
8448 SVL Box
Victorville, CA 92392

Larry Ulrich
P.O. Box 178
Trinidad, CA 95570

Rick York
California Energy Commission
c/o Environmental
Protection Division
1516 9th Street M.S. 40
Sacramento, CA 95814

Gary Zahm
1804 Crescent Court
Los Banos, CA 93635

PAGE 218. LARGE PHOTO: *Douglas's meadowfoam* (Limnanthes douglasii *ssp.* rosea), GERALD AND BUFF CORSI. • TOP TO BOTTOM, LEFT TO RIGHT, ROW 1: *Mountain ash* (Sorbus californica), MICHAEL GRAF; *Fawn lily* (Erythronium revolutum) *white variant*, JOHN GAME; *California pitcher plant* (Darlingtonia californica), NORDEN (DAN) CHEATHAM; *Pacific sedum* (Sedum spathulifolium), LARRY ULRICH. • ROW 2: *Cream cups* (Platystemon californica), MARK SKINNER. • ROW 3: *Superb Mariposa lily* (Calochortus superbus), WILLIAM T. FOLLETTE. • ROW 4: *Diogenes' lantern* (Calochortus amabilis), LARRY ULRICH; *Chinese houses* (Collinsia heterophylla), LARRY ULRICH; *Smooth yellow violet* (Viola glabella), LARRY ULRICH; *Silvery bush lupine* (Lupinus argenteus), LARRY ULRICH. • ROW 5: *California poppy* (Eschscholzia californica), CRAIG LIPPUS; *Buckwheat* (Eriogonum sp.), LARRY ULRICH; *Coast larkspur* (Delphinium decorum *ssp.* decorum), LARRY ULRICH; *Cantelow's lewisia* (Lewisia cantelovii), MARK SKINNER.

PAGE 219. LARGE PHOTO: *Giant trillium* (Trillium chloropetalum), LARRY ULRICH. • TOP TO BOTTOM, LEFT TO RIGHT, ROW 1: *Footsteps of spring* (Sanicula arctopoides), COURTESY OF CNPS; *Thistle sage* (Salvia carduacea), SUSAN COCHRANE; *California hesperochiron* (Hesperochiron californicus), WILLIAM T. FOLLETTE; *Striped adobe-lily* (Fritillary striata), JOHN GAME. • ROW 2: *California coreopsis* (Coreopsis californica), GERALD AND BUFF CORSI; *Lyall's angelica* (Angelica arguta), DON EASTMAN; *Desert mariposa lily* (Calochortus kennedyi), TOMMY DODSON; *Mule ears* (Wyethia helenoides), WALT ANDERSON. • ROW 3: *Kellogg's lily* (Lilium kelloggii), LARRY ULRICH. • ROW 4: *Bird's beak* (Cordylanthus maritimus), JOE MEDEIROS; *San Luis Obispo mariposa lily* (Calochortus obispoensis), MARK SKINNER. • ROW 5: *Monkey flower* (Mimulus guttatus), TOMMY DODSON; *Fiesta flower* (Pholistoma auritum *var.* auritum), WILLIAM T. FOLLETTE.